M336
Mathematics and Computing: a third-level course

GROUPS & GEOMETRY

UNIT IB4
GROUPS: AXIOMS AND THEIR CONSEQUENCES

Prepared for the course team by
Bob Coates

The Open University

This text forms part of an Open University third-level course.
The main printed materials for this course are as follows.

Block 1
Unit IB1   Tilings
Unit IB2   Groups: properties and examples
Unit IB3   Frieze patterns
Unit IB4   Groups: axioms and their consequences

Block 2
Unit GR1   Properties of the integers
Unit GR2   Abelian and cyclic groups
Unit GE1   Counting with groups
Unit GE2   Periodic and transitive tilings

Block 3
Unit GR3   Decomposition of Abelian groups
Unit GR4   Finite groups 1
Unit GE3   Two-dimensional lattices
Unit GE4   Wallpaper patterns

Block 4
Unit GR5   Sylow's theorems
Unit GR6   Finite groups 2
Unit GE5   Groups and solids in three dimensions
Unit GE6   Three-dimensional lattices and polyhedra

The course was produced by the following team:

Andrew Adamyk (BBC Producer)
David Asche (Author, Software and Video)
Jenny Chalmers (Publishing Editor)
Bob Coates (Author)
Sarah Crompton (Graphic Designer)
David Crowe (Author and Video)
Margaret Crowe (Course Manager)
Alison George (Graphic Artist)
Derek Goldrei (Groups Exercises and Assessment)
Fred Holroyd (Chair, Author, Video and Academic Editor)
Jack Koumi (BBC Producer)
Tim Lister (Geometry Exercises and Assessment)
Roger Lowry (Publishing Editor)
Bob Margolis (Author)
Roy Nelson (Author and Video)
Joe Rooney (Author and Video)
Peter Strain-Clark (Author and Video)
Pip Surgey (BBC Producer)

With valuable assistance from:

Maths Faculty Course Materials Production Unit
Christine Bestavachvili (Video Presenter)
Ian Brodie (Reader)
Andrew Brown (Reader)
Judith Daniels (Video Presenter)
Kathleen Gilmartin (Video Presenter)
Liz Scott (Reader)
Heidi Wilson (Reader)
Robin Wilson (Reader)

The external assessor was:
Norman Biggs (Professor of Mathematics, LSE)

The Open University, Walton Hall, Milton Keynes, MK7 6AA.

First published 1994. Reprinted 2001, 2009.

Copyright © 1994 The Open University

All rights reserved. No part of this publication may be reproduced, stored in a retrieval system or transmitted in any form or by any means, without written permission from the publisher or a licence from the Copyright Licensing Agency Limited. Details of such licences (for reprographic reproduction) may be obtained from the Copyright Licensing Agency Ltd of 90 Tottenham Court Road, London, W1P 9HE.

Edited, designed and typeset by the Open University using the Open University TeX System.

Printed in Malta by Interprint Limited.

ISBN 0 7492 2162 3

This text forms part of an Open University Third Level Course. If you would like a copy of *Studying with The Open University*, please write to the Central Enquiry Service, PO Box 200, The Open University, Walton Hall, Milton Keynes, MK7 6YZ. If you have not already enrolled on the Course and would like to buy this or other Open University material, please write to Open University Educational Enterprises Ltd, 12 Cofferidge Close, Stony Stratford, Milton Keynes, MK11 1BY, United Kingdom.

1.3

CONTENTS

Study guide	4
Introduction	5
1 The group axioms	**6**
1.1 Elementary consequences	6
1.2 Checking the axioms	10
2 Subgroups and cosets	**12**
2.1 Subgroups	12
2.2 Cosets and Lagrange's Theorem	19
3 Normal subgroups and quotient groups	**23**
3.1 Normal subgroups	23
3.2 Quotient groups	27
4 Isomorphisms and homomorphisms	**30**
4.1 Isomorphisms	30
4.2 Homomorphisms	36
5 Generators and relations (audio-tape section)	**42**
Appendix: denoting groups	49
Solutions to the exercises	51
Objectives	64
Index	64

STUDY GUIDE

Apart from Section 1 (which is relatively light), the sections of this unit are approximately equal in terms of the study time you will probably require.

Many of the exercises which appear in this unit take the form of asking you to prove some result or participate in the development of the theory. In our experience students find their first attempts at proof difficult, and there is always the temptation to look at the solution too soon. Please resist this temptation since, the more time you spend thinking about the possible ways of obtaining solutions to the exercises in this unit, the easier you will find the proofs later on. As a general guide, most of the proofs are best constructed by working from both ends towards the middle, closing the gap by careful application of the appropriate definitions, previous results and any other information given. Of course, once this is completed, you need to check that the resulting proof can be read as a logical deduction from the first line to the last, i.e. from the assumptions to the conclusions.

There is an audio programme associated with Section 5 of this unit.

There is no video programme associated with this unit.

You will not need the *Geometry Envelope* in your study of this unit.

INTRODUCTION

In this unit we continue our investigation of some of the basic properties of groups, which we started in *Unit IB2*. In that unit we based our work on the explorations of several specific groups. In this unit we are more concerned with groups in the abstract and with developing proof strategies which you will need throughout the course.

Rather than looking at particular groups and their properties, we shall derive consequences of the group axioms *without* reference to particular groups. In this way we shall therefore obtain results which are applicable to *all* groups. Despite this generality, we shall often illustrate our results with reference to specific groups. Even when we do not, you would be well advised always to place our general results within the context of the groups that you are familiar with.

It may well be that you have seen many of the results and proofs in previous courses. In this course, however, we shall place more emphasis on your being able to construct proofs of your own, whereas in the past you were rarely expected to do more than be able to follow other people's proofs.

In Section 1 we consider some of the immediate consequences of the group axioms, in particular those which involve the identity element and inverses of elements.

Sections 2 and 3 deal with methods of obtaining new groups from old. In Section 2 we consider those subsets of groups which themselves form groups, i.e. the *subgroups* of groups. In Section 3 we consider *quotient* groups. These groups are constructed in much the same way that the group $\mathbb{Z}_n$ of integers under addition modulo n may be constructed from the group $\mathbb{Z}$ of integers under addition.

Section 4 introduces *isomorphisms* and *homomorphisms*. This is the section where we address the problem of what we mean by saying that two groups are 'essentially the same' group.

In Section 5 we discuss *generators* and *relations*. The generators are a set of elements from which we may construct *all* the elements of the group, given only the fact that the resulting elements must satisfy *both* the group axioms *and* some additional conditions (for example, commutativity) which we impose. These additional conditions are known as the relations. The concept of a group being defined in terms of generators and relations is of central importance in many subsequent units, of both the Groups and the Geometry streams.

The appendix summarizes the various notations used for groups in the first block of the course.

1 THE GROUP AXIOMS

In this section we shall derive, or ask you to derive, some of the results which follow directly from the group axioms. As we said in the Introduction, the virtue of this procedure is that, since no specific group is mentioned, it will follow that each of the results obtained is true for *all* groups.

First, we remind you of the group axioms themselves.

Definition 1.1 Group axioms

A set G together with a binary operation $\circ$ defined on G is a **group** if the following axioms are satisfied.

Closure For all pairs x, y of elements in G, $x \circ y$ is also in G.

Associativity For all choices of three elements x, y, z in G, we have
$$(x \circ y) \circ z = x \circ (y \circ z).$$

Identity There is an **identity element** e, say, in G with the property that, for any element x in G,
$$e \circ x = x = x \circ e.$$

Inverses For each element x in G, there is an **inverse element** x^{-1} in G with the property that
$$x \circ x^{-1} = e = x^{-1} \circ x,$$
where e is the identity element of G.

The set G is sometimes called the *underlying set* of the group. As usual, we shall often refer to the group itself as G despite the fact that the group is strictly the set G together with the operation $\circ$ and formally denoted by $(G, \circ)$.

As we said in *Unit IB2*, for abstract groups we shall normally omit the group operation $\circ$ and write, for example, xy instead of $x \circ y$. On the other hand, when dealing with specific groups, we shall often include the operation, as, for example, with the group $(\mathbb{Z}, +)$ of integers under addition. We may also find it helpful to include the operations when there is more than one group under consideration and we wish to distinguish between the operations of the groups.

1.1 Elementary consequences

As was observed in *Unit IB2*, though the identity axiom only guarantees 'an identity element', in the inverses axiom this has become '*the* identity element'. This implies that, as a consequence of the group axioms, there can be only one identity (in other words the guaranteed identity is *unique*).

There is a standard way of proving that an object is unique. We assume that there are two objects satisfying the conditions which specify (i.e. define) the object. Then we show that those two objects are in fact the same.

For uniqueness, assume two and show they are the same.

Theorem 1.1 Uniqueness of identity

Let G be a group. Then the identity element of G is unique.

Proof

Let the elements e and e' of G be identities. In other words e and e' both satisfy the identity axiom, that is

$$ex = x = xe \tag{1.1}$$

and

$$e'x = x = xe' \tag{1.2}$$

for *any* x in G.

We use the group axioms to show that e and e' are the same. To do this we require statements containing only e and e'. One way of obtaining such statements is to substitute e and/or e' for the general element x of G in each of Statements 1.1 and 1.2.

Take special cases.

It turns out to be most useful to take the special case $x = e'$ in Statement 1.1 and $x = e$ in Statement 1.2. When we do this we obtain

$$ee' = e' = e'e \qquad (1.3)$$

and

$$e'e = e = ee'. \qquad (1.4)$$

From Statement 1.3 we see that $ee' = e'$, and from Statement 1.4 that $ee' = e$. Therefore e and e' are the same (since both are equal to ee'). So the group G has a unique identity. ∎

The second uniqueness result we wish to obtain is the uniqueness of the inverse of a given element.

Theorem 1.2 Uniqueness of inverses

Let G be a group and a be an element of G. Then the inverse of a is unique.

Exercise 1.1

Prove Theorem 1.2, carefully justifying each step of your argument by reference to the group axioms (including the associativity axiom).

(Note that we ask you to use the associativity axiom here to help you convince yourself that you appreciate where it is needed. We shall stop using this axiom explicitly in proofs shortly.)

Hint Assume that x and y are both inverses for a and show that they are the same.

The closure axiom tells us that we may form the product of any *two* elements using the given operation, and also that the result will be another element of the group. When it comes to forming the product of three elements, x, y and z, *in that order*, using the given binary operation, there are just two possible ways in which this can be done. One way is to calculate the product of x and y and then form the product of the result with z. This produces the product $(xy)z$. The other way is to find the product of x with the product of y and z, giving $x(yz)$. What the associativity axiom tells us is that these two products are the same.

Remember that the order of the elements is crucial since, for a general group, it may not be true that $xy = yx$. Just consider, for example, the case of D_6, where
$$sr = r^5s \neq rs,$$
or the products of the two matrices
$$\begin{bmatrix} 1 & 0 \\ 0 & -1 \end{bmatrix} \text{ and } \begin{bmatrix} 0 & 1 \\ 1 & 0 \end{bmatrix}.$$

Exercise 1.2

Write down the five ways that we can form the product of the four elements w, x, y and z, *in that order*.

In fact, as you might hope, all five products produce the same answer. For example,

$$w(x(yz)) = w((xy)z) \quad \text{(by the associativity of } x, y \text{ and } z\text{)}$$
$$= (w(xy))z \quad \text{(by the associativity of } w, (xy) \text{ and } z\text{)}.$$

Exercise 1.3

Prove, using the associativity axiom, that each of the remaining products of four elements produces the same result.

With some effort the method used above could be extended to products of five (or more) elements. In fact, as we remarked in *Unit IB2*, it is possible, using the Principle of Mathematical Induction, to prove the following result.

> ***Theorem 1.3 Generalized associativity rule***
>
> Let G be a group and n be a positive integer.
> Then the product of n elements of G, *taken in a fixed order*, is the same no matter in which way the product is calculated.

The proof of this theorem is more involved than illuminating, so we have omitted it from the course. We shall however assume this result and write such expressions without brackets, doing the calculation by combining adjacent pairs of elements in the most convenient way.

We now establish a result which is very helpful when performing calculations with elements of groups, and which also has several useful consequences.

> ***Theorem 1.4 Cancellation rule***
>
> Let G be a group and let x, y and a be elements of G. Then
>
> $$ax = ay \quad \text{implies} \quad x = y.$$

Proof

Since G is a group and a is an element of G, by the inverses axiom it has an inverse element a^{-1} in G.

We are given that $ax = ay$. Thus, when we multiply both sides of this equation on the left by a^{-1}, we obtain

$$a^{-1}(ax) = a^{-1}(ay).$$

Using the associativity axiom gives

$$(a^{-1}a)x = (a^{-1}a)y.$$

By the inverses axiom this becomes

$$ex = ey,$$

where e is the identity of the group.

Lastly, by the identity axiom, we have the required result that

$$x = y. \qquad \blacksquare$$

The proof of Theorem 1.4 contains absolutely all the details which would be required by even the most particular examiner. (We even made explicit use of associativity axiom, which we promised to stop doing shortly.) As the course progresses we shall include fewer details in proofs. Nevertheless, when you read our proofs, and when you construct proofs for yourself, you must always be sure that you could justify each step — the justification being a reference either to an axiom or to an earlier result.

Strictly speaking we should have referred to the above result as the *Left* Cancellation Rule, since it tells us that we may cancel the element a from the *left*-hand side of the equation $ax = ay$.

Exercise 1.4

State and prove the *Right* Cancellation Rule, giving the same amount of detail as we gave in the proof of the Left Cancellation Rule.

A nice application of the cancellation rules is to determine the inverse of the product of two elements of a group. The result is as follows.

> **Theorem 1.5 Inverse of a product**
>
> Let G be a group and x and y be elements of G.
> Then the inverse of the element xy is $y^{-1}x^{-1}$.

In other words,
$$(xy)^{-1} = y^{-1}x^{-1}.$$
The fact that there is a reversal of order in the inverse of a product may seem strange initially. However, it is quite clear that the reverse process of putting on socks and then putting on shoes certainly involves taking off the shoes first.

Proof

By the inverses axiom, we know that $(xy)(xy)^{-1} = e$, where e is the identity of the group. If we could show that $(xy)(y^{-1}x^{-1}) = (xy)(xy)^{-1}$, then the Left Cancellation Rule would produce the desired result.

Therefore, we consider the product $(xy)(y^{-1}x^{-1})$.

$$\begin{aligned}(xy)(y^{-1}x^{-1}) &= x((yy^{-1})x^{-1}) \quad \text{(by the associativity axiom)} \\ &= x(ex^{-1}) \quad \text{(by the inverses axiom)} \\ &= xx^{-1} \quad \text{(by the identity axiom)} \\ &= e \quad \text{(by the inverses axiom)}.\end{aligned}$$

However, as we observed at the start of the proof, $(xy)(xy)^{-1} = e$. So

$$(xy)(y^{-1}x^{-1}) = e = (xy)(xy)^{-1}.$$

Therefore, using the Left Cancellation Rule,

$$y^{-1}x^{-1} = (xy)^{-1}. \qquad \blacksquare$$

Exercise 1.5

Give an alternative proof of Theorem 1.5 using the uniqueness of inverse elements.

Exercise 1.6

Show that $\left(x^{-1}\right)^{-1} = x$.

Theorem 1.5 has an obvious generalization to the inverse of a product of any number of elements, namely

$$(x_1 x_2 \ldots x_n)^{-1} = x_n^{-1} \ldots x_2^{-1} x_1^{-1}.$$

We shall use this result when required but will spare you the details of the proof.

As you may suspect, this result is proved using the Principle of Mathematical Induction.

The solution to Exercise 1.6 might have made you wonder whether you need to show that an element behaves as both the right *and* left inverse of an element in order to be able to say that it *is* the inverse.

> **Lemma 1.1**
>
> Let G be a group with identity e and let x be an element of G.
> If $xy = e$ for some element y of G, i.e. y behaves like a right inverse for x, then $y = x^{-1}$, i.e. it *is* the inverse of x.

Proof

We are given that $xy = e$. Furthermore $xx^{-1} = e$, by the inverses axiom. Equating the two expressions for e gives

$$xy = xx^{-1}.$$

Using the Left Cancellation Rule, we have $y = x^{-1}$. ∎

An alternative proof involves multiplying both sides of the equation $xy = e$ on the left by the element x^{-1}.

Exercise 1.7

Let G be a group with identity e and let x be an element of G. Show that, if $yx = e$ for some element y of G, i.e. y behaves like a left inverse for the element x, then $y = x^{-1}$.

The solution to Exercise 1.7 completes the proof of the following theorem.

Theorem 1.6

Let G be a group with identity e and let x be an element of G. If y is an element of G such that *either* $yx = e$ *or* $xy = e$, then $y = x^{-1}$.

If y behaves as *either* a left inverse *or* a right inverse of x, then y is *the* inverse of x.

Checking for an identity simplifies in a similar manner.

Exercise 1.8

Let G be a group with identity e and let x be an element of G.

(a) Show that, if f is an element of G such that $fx = x$, then $f = e$. In other words, show that, if f behaves like a left identity for the particular element x of G, then f *is* the identity of the group.

(b) State and prove the corresponding result with 'left' replaced by 'right'.

The solution to Exercise 1.8 provides us with a proof of the following theorem.

Theorem 1.7

Let G be a group with identity e and let x be an element of G. If f is an element of G such that *either* $fx = x$ *or* $xf = x$, then $f = e$.

1.2 Checking the axioms

We now introduce a method of constructing a new group from two existing ones: the group constructed is called the *direct product* of the two original groups. This method of construction will be of central importance in the Groups stream of the course, and will be used here to give you practice at checking the group axioms for particular cases.

The construction starts with two groups, $(G, \circ)$ and $(H, *)$. To define the new group, we first define its underlying set and then define the operation on that set.

Example 1.1

To illustrate the construction of the direct product of two groups, we look at an example that you have already met: the plane represented by the usual coordinate system.

Each point of the plane is represented by an *ordered pair* (x, y) of real numbers. In other words, the coordinates of points of the plane are elements of the *Cartesian product*

$$\mathbb{R} \times \mathbb{R} = \{(x, y) : x \in \mathbb{R}, y \in \mathbb{R}\}.$$

The Cartesian product $\mathbb{R} \times \mathbb{R}$ is also denoted by $\mathbb{R}^2$.

The operation of vector addition on $\mathbb{R} \times \mathbb{R}$ is defined 'component-wise' in terms of the operation of addition on $\mathbb{R}$ as follows:

$$(x_1, y_1) + (x_2, y_2) = (x_1 + x_2, y_1 + y_2). \qquad \blacklozenge$$

We can generalize the definitions of both the set and the operation as follows.

Given two groups $(G, \circ)$ and $(H, *)$, we define the underlying set of the new group to be the **Cartesian product** of the sets G and H, that is the set

$$G \times H = \{(g, h) : g \in G, h \in H\}.$$

In Example 1.1, $G = H = \mathbb{R}$.

As with $\mathbb{R} \times \mathbb{R}$, the new operation is defined by applying the original operations to the separate components. Generalizing requires a little care because the operations for the two component groups may be different. As first components come from G, with operation $\circ$, and second components come from H, with operation $*$, the general definition of the new operation (which we call $\square$) is defined by

$$(g_1, h_1) \,\square\, (g_2, h_2) = (g_1 \circ g_2, h_1 * h_2).$$

Definition 1.2 Direct product of two groups

Let $(G, \circ)$ and $(H, *)$ be two groups. Their **direct product** $(G \times H, \square)$ is the group defined as follows.

Set The underlying set of the group is $G \times H$, the Cartesian product of the underlying sets G and H,

$$G \times H = \{(g, h) : g \in G, h \in H\}.$$

Operation If (g_1, h_1) and (g_2, h_2) are elements of $G \times H$ then

$$(g_1, h_1) \,\square\, (g_2, h_2) = (g_1 \circ g_2, h_1 * h_2).$$

Taking on trust for the moment the fact that this definition does produce a group, the following exercise is designed to give you some 'hands on' experience of direct products of groups.

Exercise 1.9

Write out the Cayley table of the group $\mathbb{Z}_2 \times \mathbb{Z}_3$. Use $+$ for the new operation.

We now establish that the direct product of two groups always is a group. To get a feel for how the proof will go, we ask you to verify that $\mathbb{R} \times \mathbb{R}$, with the operation of vector addition defined above, is a group.

Exercise 1.10

Prove that the set $\mathbb{R} \times \mathbb{R}$ with the operation of vector addition is a group.

The proof for the general case follows the solution to Exercise 1.10 closely.

To verify that $(G \times H, \square)$ is a group, we must check the four group axioms.

To check that the above definition of $\square$ does indeed produce a closed binary operation on $G \times H$ (in other words, it does satisfy the closure axiom), we observe the following.

Let (g_1, h_1) and (g_2, h_2) be any two elements of $G \times H$. By the definition of $G \times H$, we know that g_1 and g_2 are elements of G and that h_1 and h_2 are elements of H.

Since G is a group, it is closed under the operation $\circ$, therefore $g_1 \circ g_2$ is in G. Since H is a group, it is closed under the operation $*$, so $h_1 * h_2$ is in H. Therefore, by the definition of $\square$, it follows that

$$(g_1, h_1) \square (g_2, h_2) = (g_1 \circ g_2, h_1 * h_2) \in G \times H.$$

In the following exercise we ask you to check the remaining group axioms for $(G \times H, \square)$.

Exercise 1.11

(a) Prove that $\square$ is associative (that is, check the associativity axiom).

(b) Let the identity of G be e and the identity of H be f. Prove that (e, f) is the identity of $G \times H$ (that is, check the identity axiom).

(c) Let (g, h) be an element of $G \times H$, where g has inverse g^{-1} in G and h has inverse h^{-1} in H. Prove that (g^{-1}, h^{-1}) is the inverse of (g, h) in $G \times H$ (that is, check the inverses axiom).

2 SUBGROUPS AND COSETS

2.1 Subgroups

In Section 2 of *Unit IB2*, we defined a subgroup of a group and introduced you to some specific examples.

Definition 2.1 Subgroup

A **subgroup** H of a group G is a subset H of G which is itself a group under the group operation of G.

We use the notation

$$H \leq G$$

to denote that H is a subgroup of G.

We saw in *Unit IB2* that if H is a subset of G then, since *all* elements of G obey the associativity axiom, all elements of H must also obey this axiom (once we have verified that H is closed). So checking the subgroup axioms for a subset H of a group G reduces to the following.

Definition 2.2 Subgroup axioms

If G is a group with binary operation $\circ$ and H is a subset of G, then H is a **subgroup** of G (with binary operation $\circ$) provided that the following axioms are satisfied.

Closure For all pairs x, y of elements in H, $x \circ y$ is also in H.

Identity The identity element e of G is in H.

Inverses For each element x in H, the inverse element x^{-1} in G is also in H.

Note that, if $x, y \in H$ and $x = e$, then $xy = ey = y \in H$. Similarly, if $y = e$, then $xy = x \in H$. Therefore, in checking for closure, we need only check pairs x, y neither of which is equal to the identity.

Exercise 2.1

Prove that the set $\{e, (123), (132)\}$ is a subgroup of the permutation group S_3.

There are several alternative ways of testing whether a particular subset of a group is a subgroup. The one we give in the following theorem is useful in practice because it is somewhat simpler to apply than checking the three subgroup axioms.

> **Theorem 2.1**
>
> Let G be a group and H be a subset of G.
> Then H is a subgroup of G if and only if it satisfies the following two properties:
> (a) H is non-empty, i.e. $H \neq \varnothing$;
> (b) $x, y \in H$ implies $x^{-1}y \in H$.

Proof

The 'if' part of the statement of the theorem says that

if H has the two properties stated, then H is a subgroup of G.

We are given that H is a non-empty subset of G and that, if x and y are in G, then $x^{-1}y$ is in G. We have to show that H is a subgroup of G; in other words, that it satisfies the three subgroup axioms. It turns out that the easiest way of proving them is not in the order in which we listed the axioms.

Identity

By the first property, H is non-empty. Therefore there exists some element x in H.
By the second property, x and y in H imply that $x^{-1}y$ is in H. Taking the special case $y = x$ shows that $x^{-1}x = e$ is in H.
Thus H satisfies the identity axiom.

Inverses

If x is in H then, by the identity axiom (already verified), both x and e are in H.
It follows therefore, by the second property, that $x^{-1}e = x^{-1}$ is in H.
So H satisfies the inverses axiom.

Closure

Let x and y be elements of H. By the inverses axiom (already verified), x^{-1} is an element of H. Therefore x^{-1} and y are both in H.
So, by the second property, $\left(x^{-1}\right)^{-1} y = xy$ is in H.
Therefore H satisfies the closure axiom.

So, we have proved the 'if' part of the theorem.

The 'only if' part of the statement of theorem says that

if H is a subgroup of G, then H has the two properties stated.

We ask you to prove the 'only if' part of the theorem in the following exercise. □

Exercise 2.2

Let G be a group and H be a subgroup of G.
Prove that H is a non-empty subset of G and that, if x and y are elements of H, then so is $x^{-1}y$.

Proof of Theorem 2.1 continued

So, we have proved both parts of the theorem. ■

Exercise 2.3

Prove the following alternative subgroup test.
Let G be a group and H be a subset of G. Then H is a subgroup of G if and only if it satisfies the following two properties:

(a) H is non-empty, i.e. $H \neq \emptyset$;

(b) $x, y \in H$ implies $xy^{-1} \in H$.

We now have two alternative tests to use to decide whether a subset of a group is a subgroup. Written in additive notation these are as follows.

> A subset H of an additive group G is a subgroup if and only if it satisfies the conditions:
>
> (a) H is non-empty;
>
> (b) $x, y \in H$ implies $-x + y \in H$.

> A subset H of an additive group G is a subgroup if and only if it satisfies the conditions:
>
> (a) H is non-empty;
>
> (b) $x, y \in H$ implies $x - y \in H$.

The second form of the conditions looks rather more natural, and is the version we shall generally use for Abelian groups. We illustrate this result in a specific case by using it to prove that $(2\mathbb{Z}, +)$ is a subgroup of the group $(\mathbb{Z}, +)$.

$2\mathbb{Z} = \{2n : n \in \mathbb{Z}\}$.

Firstly, $2\mathbb{Z}$ is non-empty since, for example, $0 = 2 \times 0$ is in $2\mathbb{Z}$.

Secondly, let x and y be two elements of $2\mathbb{Z}$. By the definition of $2\mathbb{Z}$, there exist integers m and n such that $x = 2m$ and $y = 2n$. So

$$x - y = 2m - 2n = 2(m - n).$$

However $(\mathbb{Z}, +)$ is a group, so $m - n$ is an element of $\mathbb{Z}$. Therefore $2(m - n)$ is an element of $2\mathbb{Z}$.

This completes the proof.

As a further illustration of the use of Theorem 2.1, we prove the following.

> **Theorem 2.2**
>
> Let G be a group and H_1 and H_2 be subgroups of G.
> Then the intersection $H_1 \cap H_2$ is a subgroup of G (and also a subgroup of both H_1 and H_2).

Proof

Firstly, since H_1 and H_2 are subgroups of G, each contains the identity e of G. Therefore e is an element of their intersection.
Hence the intersection is non-empty.

Secondly, if x and y are in the intersection, then x and y must be in both H_1 and H_2. By Theorem 2.1, $x^{-1}y$ is in both H_1 and H_2.
Therefore $x^{-1}y$ is in the intersection.

Thus we have verified that the intersection $H_1 \cap H_2$ has the two properties of the theorem, and so $H_1 \cap H_2 \leq G$.

Since, by definition, $H_1 \cap H_2$ is a subset of both of the subgroups H_1 and H_2, and since, as we have proved, it is a subgroup of G, we also have $H_1 \cap H_2 \leq H_1$ and $H_1 \cap H_2 \leq H_2$. ∎

The above theorem generalizes to any (non-empty) collection of subgroups — their intersection is always a subgroup. For a finite collection, the proof is almost identical to that above. The result is still true for infinite collections of subgroups, though the proof in this case is a little more complicated because of some notational difficulties. Neither proof is given here, though we shall use the results later. For convenience, we state the general result.

Theorem 2.3

Let G be a group. The intersection of any non-empty collection of subgroups of G is a subgroup both of G and of each of the subgroups in the collection.

A natural question which arises now is whether the union of subgroups of a given group is also a subgroup. The answer is *no*.

It is a fact of life that, in algebra, intersections are always much nicer than unions.

Exercise 2.4

Give an example of a group with two subgroups such that their union is not a subgroup.

Hint An easy example can be found by considering the group $\mathbb{Z}_6$ under addition modulo 6.

In the remainder of this subsection we shall show how Theorem 2.3 can be used to define the phrase

   the subgroup generated by the set of elements ...

which we introduced informally in Section 3 of *Unit IB2*.

If the generating set contains just one element, say x, then the subgroup generated is a cyclic group denoted by $\langle x \rangle$. In addition, we are able to describe precisely what its elements are:

$$\langle x \rangle = \{x^n : n \in \mathbb{Z}\}.$$

For generating sets with more than one element the situation becomes slightly more complicated and much more interesting. In what follows, we shall deal with the problem of whether such subgroups exist and what their elements look like. First of all we give the formal definition.

Definition 2.3 Subgroup generated by a set

Let G be a group and S be a subset of (the underlying set of) G.
We define the **subgroup of G generated by S** to be the smallest subgroup of G containing the set S.

We denote this subgroup by $\langle S \rangle$.

By 'smallest' in Definition 2.3 we mean that it is a subgroup of G containing S and that it is contained in every other subgroup which contains S.

The notation $\langle S \rangle$ is a generalization of the notation $\langle x \rangle$ used for cyclic subgroups generated by the single element x. To be consistent, we ought to change the previous notation to $\langle \{x\} \rangle$, though, like everybody else, we shall not do so.

The fact that such a subgroup exists (and is unique) is a consequence of
Theorem 2.3.

Theorem 2.4

Let G be a group and S be a subset of (the underlying set of) G.
Then there exists a unique smallest subgroup of G which contains S.

In other words, S generates a unique subgroup of G.

Proof

Existence

Consider the collection of all subgroups of G which contain the set S.

This collection is certainly non-empty, since it contains the subgroup G.

We saw in *Unit IB2* that every group is a subgroup of itself.

Now, by Theorem 2.3, the intersection H of all of these subgroups is itself a subgroup of G.
Furthermore, since each of the subgroups in the collection contains the set S, so does their intersection H.
So H is a subgroup of G containing S.

In addition, if any subgroup of G contains the set S, it is in our collection and so must contain the intersection H.

So H is a smallest subgroup of G containing S.

Uniqueness

Suppose there were two such smallest subgroups containing S. Then, by the definition of 'smallest', each would be a subset of the other. Hence the two subgroups would be equal. Therefore the subgroup generated by S is unique. ∎

Theorem 2.4 tells us that the subgroup $\langle S \rangle$ exists and is unique. Provided that we know about *all* the subgroups of G, it also tells us how to construct $\langle S \rangle$ (i.e. by intersection). However, it does not tell us what the elements of $\langle S \rangle$ look like.

There is a method of constructing $\langle S \rangle$ which rectifies this deficiency. This method takes a 'hands on' approach. The details of this construction are similar to other important constructions which we shall meet in Sections 3 and 5 of this unit.

Example 2.1

To get a feel for what the general situation looks like, let us construct the subgroup of the group D_6 generated by the set $\{r^2, s\}$, i.e. the subgroup $\langle r^2, s \rangle$. We do so by investigating the consequences of demanding that $\langle r^2, s \rangle$ should be a subgroup which contains r^2 and s. As a reminder, when we described D_6 as the group of symmetries of the regular hexagon, r was rotation through $\pi/3$ and s was a reflection in an axis of symmetry.

As with single generators, we omit the set notation $\{\ldots\}$ when we have a finite number of generators.

A more informal approach to constructing the subgroup of D_6 generated by r^2 and s was considered in *Unit IB2*.

Every subgroup must contain the identity element, e, therefore $\langle r^2, s \rangle$ contains e, r^2 and s.

Every subgroup must contain the inverses of its elements, so $\langle r^2, s \rangle$ must certainly contain the inverses of the generators,

$$(r^2)^{-1} = r^4 \quad \text{and} \quad s^{-1} = s.$$

Thus,

$$\{e, r^2, r^4, s\} \subseteq \langle r^2, s \rangle.$$

We have not yet investigated the consequences of requiring closure. This tells us that we must include all possible products of the elements obtained so far.

So what do these new products actually give? We can omit any occurrence of e from a product without changing it. We therefore need to consider only products of the form

$$x_1 x_2 \ldots x_n,$$

where each x_i is either an even power of r or is s. ◊

Exercise 2.5

Use the relations $sr = r^5 s$ and $r^6 = e$ to show that, if k is a positive integer, then

$$sr^{2k} = r^{2l}s,$$

for some integer l. Show also that we need only consider $l = 0, 1, 2$.

Example 2.1 continued

The result of the last exercise means that if we start with a product of the form

$$x_1 x_2 \ldots x_n,$$

where each x_i is either an even power of r or is s, and move all the occurrences of r to the front, we obtain one of the standard forms

$$r^{2l}, \quad l = 0, 1, 2,$$
$$r^{2l}s, \quad l = 0, 1, 2,$$

according to whether the original product contained an even or an odd number of occurrences of s.

We now know that

$$\{e, r^2, r^4, s, r^2 s, r^4 s\} \subseteq \langle r^2, s \rangle.$$

Further, the above set is closed, since we know that any product of occurrences of s and even powers of r can be reduced to one of the above standard forms.

The above set also contains an inverse for each of its elements. We could check this directly, or note that each element can be expressed as a product of the original generators,

$$r^2, s,$$

and their inverses,

$$r^4, s,$$

and so its inverse can be expressed similarly. For example,

$$\begin{aligned}(r^2 s)^{-1} &= s^{-1}(r^2)^{-1} \\ &= sr^4 \\ &= r^2 s \quad \text{(using } sr = r^5 s \text{ four times and } r^6 = e \text{ three times).}\end{aligned}$$

It follows that

$$\{e, r^2, r^4, s, r^2 s, r^4 s\}$$

is a subgroup of D_6 containing r^2 and s and contained in $\langle r^2, s \rangle$. Hence, by the 'smallest' property,

$$\{e, r^2, r^4, s, r^2 s, r^4 s\} = \langle r^2, s \rangle.$$ ♦

The whole construction, from generators, adding the identity, including inverses of generators and, finally, taking all products, works quite generally and produces a subgroup. The closure relies on products of products being products. The existence of inverses depends on the inverse of a product being the product of inverses in reverse order.

The only part of the example that cannot always be done is expressing the elements in some 'tidy' standard form.

We now carry out the general version of this construction.

Let G be a group and S be a subset of G. We wish to describe the elements of the subgroup $\langle S \rangle$ of G generated by the set S.

By the identity axiom, *any* subgroup must contain the identity element e of G.

By the inverses axiom, any subgroup containing the set S also contains the element s^{-1} for every s in S.

These two facts mean that the subgroup we are looking for must certainly contain the union

$$S \cup \{e\} \cup \{s^{-1} : s \in S\}.$$

We shall denote this union by $\hat{S}$.

From what we have just said, the subgroup generated by S is the same as the subgroup generated by $\hat{S}$.

At this stage we know that $\hat{S}$ contains all elements of S and the identity element e. In addition, every element in $\hat{S}$ has an inverse in $\hat{S}$. However, we do not know that $\hat{S}$ is a group, because it may not satisfy the closure axiom.

By the closure axiom, since $\langle S \rangle$ is the subgroup generated by $\hat{S}$, it must also contain all products of any finite number of elements of $\hat{S}$.

This prompts the following definition.

> We take the 'product' of one element to be the element itself. Some other texts take the 'product' of no elements of a group to be the identity element of the group.

Definition 2.4 Words

Let G be a group and A be a subset of G. By the set of **words** in elements of A we mean the set of all products of any finite number of elements of A.

We denote this set of words by $W(A)$. That is,

$$W(A) = \{a_1 a_2 \cdots a_n : n \in \mathbb{N}, \text{ where } a_i \in A, \ 1 \leq i \leq n\}.$$

> $\mathbb{N}$ is the set of positive integers, i.e. $\mathbb{N} = \{1, 2, 3, \ldots\}$.

So the set $W(\hat{S})$ denotes the set of words in the elements of $\hat{S}$.

From what we have said about closure,

$$W(\hat{S}) \subseteq \langle S \rangle.$$

What we now need to verify is that our construction has gone far enough and that $W(\hat{S}) = \langle S \rangle$. To do so we require the other set inclusion, namely

$$\langle S \rangle \subseteq W(\hat{S}).$$

In practice, what we do is to prove that $W(\hat{S})$ is a subgroup of G. The required result then follows by the minimality of $\langle S \rangle$.

We summarize the construction above in the following theorem, and complete what remains of the proof as outlined.

Theorem 2.5

Let G be a group and S be a subset of G. Define

$$\hat{S} = S \cup \{e\} \cup \{s^{-1} : s \in S\}.$$

Then the set of all words in elements of $\hat{S}$,

$$W(\hat{S}) = \{s_1 s_2 \cdots s_n : n \in \mathbb{N}, \text{ where } s_i \in \hat{S}, \ 1 \leq i \leq n\},$$

is the subgroup of G generated by S.

Proof

From the comments preceding the statement of the theorem, we know that we need only verify that $W(\hat{S})$ is a subgroup of G.

Closure

If x and y are both in $W(\hat{S})$, that is they are both products of a finite number of elements of $\hat{S}$, then their product xy is also a product of a finite number of elements of $\hat{S}$, and so is in $W(\hat{S})$.

Identity

From the construction of $\hat{S}$, we know that it contains e. Therefore e, being a one-element product, is an element of $W(\hat{S})$.

Inverses

Let $x = s_1 s_2 \cdots s_n$ be any element of $W(\hat{S})$. Since each element s_i is in $\hat{S}$, it follows that each s_i^{-1} is in $\hat{S}$. Therefore the element $x^{-1} = s_n^{-1} \cdots s_2^{-1} s_1^{-1}$ is also in $W(\hat{S})$. ∎

In fact, with virtually no modification, the proof of the last theorem can be used to show a little more.

> **Corollary 2.1**
>
> Let G be a group and let T be a non-empty subset of G which is closed under taking inverse elements.
> Then the set $W(T)$, of all words in elements of T, is a subgroup of G.

The reason that this result is true is that T has the same properties as $\hat{S}$ above.

2.2 Cosets and Lagrange's Theorem

In your previous mathematical studies you should have encountered the concept of a *left coset*.
We remind you of the definition.

> **Definition 2.5 Left coset**
>
> Let G be a group, H be a subgroup of G and a be an element of G.
> We define the **left coset** of H by a, which we denote by aH, to be the set
>
> $aH = \{ah : h \in H\}$.

In general, for two subsets A and B of G, we define the *product* $AB = \{ab : a \in A, b \in B\}$.
So, with this convention, the left coset aH is defined to be the product $\{a\}H$.

Exercise 2.6

Write down all the left cosets of the subgroup $\{e, (12)\}$ in the permutation group S_3.

So, for each element a and subgroup H of a group G, we have defined a left coset aH.

Example 2.2

In the solution to Exercise 2.6, with $G = S_3$ and $H = \{e, (12)\}$, we saw that the two left cosets eH and $(12)H$ are the same, as are $(13)H$ and $(123)H$, and indeed $(23)H$ and $(132)H$.

So, in general, for different elements a and b of G, the left cosets aH and bH may turn out to be the same.

In this example, too, the distinct cosets $\{e, (12)\}$, $\{(13), (123)\}$ and $\{(23), (132)\}$ have a union which is the whole group. They are also disjoint, i.e. the intersection of any two is the empty set. Furthermore, each coset has the same number of elements.

These results generalize and lead to Lagrange's Theorem. ♦

We shall prove Lagrange's Theorem shortly, but first we look at the conditions under which two cosets are the same.

From our particular example, it seems possible that the coset aH is equal to H precisely when (i.e. if and only if) the element a is in H.

Exercise 2.7

Verify that the above conjecture is true when G is the group

$$D_6 = \{e = r^0, r, r^2, r^3, r^4, r^5, s, rs, r^2s, r^3s, r^4s, r^5s\}$$

and H is the subgroup

$$\langle r^2 \rangle = \{e, r^2, r^4\}.$$

We now prove the general result.

Lemma 2.1

Let G be a group and H be a subgroup of G.
Then $aH = H$ if and only if a is an element of H.

Proof

If

We are given that a is an element of the subgroup H, and we want to prove that $aH = H$.

Firstly we show that $aH \subseteq H$.
Let x be an element of aH. Then $x = ah$, for some h in H. Now a and h are elements of H and so, by the closure axiom, $ah = x$ is an element of H.
Therefore aH is a subset of H.

Secondly we show that $H \subseteq aH$.
Let y be an element of H. We may write $y = a(a^{-1}y)$ and, as both a and y are in H, so is $a^{-1}y$, by Theorem 2.1. So $y = a(a^{-1}y)$ is an element of aH.
Therefore H is a subset of aH.

Hence $aH = H$.

Only if

Here we are given that $aH = H$, and we wish to show that a is an element of H.

Since H is a subgroup, the identity element e is in H, therefore ae is in aH. However, $ae = a$ and $aH = H$, so a is in H. ∎

Since $H = eH$, we have proved a result about the equality of the cosets aH and eH. A more general result concerning the equality of cosets is given in the following theorem.

Theorem 2.6

Let G be a group, H be a subgroup of G and a and b be elements of G. Then the left cosets aH and bH are equal if and only if $a^{-1}b \in H$.

Proof

If

We are given that $a^{-1}b$ is an element of H, so $a^{-1}b = h$ for some h in H. To show that $aH = bH$, we must show that $aH \subseteq bH$ and that $bH \subseteq aH$.

Firstly we show that $aH \subseteq bH$.
Let x be an element of aH, so that $x = ah_1$ for some h_1 in H. From $a^{-1}b = h$ we deduce (by multiplying both sides on the left by a and on the right by h^{-1}) that $a = bh^{-1}$, and so $x = bh^{-1}h_1$.
But $h^{-1}h_1$ is an element of H. So x is an element of bH.
Therefore aH is a subset of bH.

Secondly we show that $bH \subseteq aH$.
Let y be an element of bH, so that $y = bh_2$ for some h_2 in H. From $a^{-1}b = h$ we deduce (by multiplying both sides on the left by a) that $b = ah$, and so $y = ahh_2$.
But hh_2 is an element of H. So y is an element of aH.
Therefore bH is a subset of aH.

Only if

We ask you to prove this part of the theorem in the following exercise. □

Exercise 2.8

Prove the 'only if' part of Theorem 2.6.

Proof of Theorem 2.6 continued

So, we have proved both parts of the theorem. ∎

Since the statement $aH = bH$ in Theorem 2.6 is symmetric in a and b, we can deduce that, in addition,

$aH = bH$ if and only if $b^{-1}a \in H$.

Theorem 2.6 also provides a short proof of the following.

Theorem 2.7

Let G be a group, H be a subgroup of G and a and b be elements of G. Then the left cosets aH and bH are either disjoint or equal.

Two sets are *disjoint* if they have no elements in common.

Proof

Since the cosets aH and bH are either disjoint or not, we need to show that if the left cosets aH and bH are not disjoint then they are equal.

Let us assume therefore that they are not disjoint, so that there exists some element x which is in both aH and bH.
From the definition of left cosets, this means that $x = ah_1$ and $x = bh_2$ for some h_1 and h_2 in H.
Therefore $ah_1 = bh_2$. Multiplying on the left by a^{-1} and on the right by h_2^{-1} gives $a^{-1}b = h_1 h_2^{-1}$.
As $h_1 h_2^{-1}$ is an element of H, Theorem 2.6 tells us that $aH = bH$. ∎

We now need one more preliminary result before our proof of Lagrange's Theorem. This is the result which says that all cosets have the same number of elements.

Exercise 2.9

Let G be a group, H be a subgroup of G and a and b be elements of G.

(a) Prove that the function

$$f : H \to aH$$
$$h \mapsto ah$$

is both one–one and onto, and hence that the sets H and aH have the same number of elements.

(b) Deduce that the left cosets aH and bH have the same number of elements.

Two sets A and B have the same number of elements if and only if there is a one–one and onto function (a bijection) from A to B. This is true even when the two sets are infinite, as the existence of such a function is the definition of 'having the same number of elements'.

We now just need a couple of definitions before we can state and prove Lagrange's Theorem.

Definition 2.6 Finite and infinite groups

A **finite group** is one whose underlying set is finite, i.e. has a finite number of elements. If the underlying set is infinite, the group is an **infinite group**.

Definition 2.7 Order of group

The **order** of a finite group G is the number of elements it contains, and is denoted by $|G|$. An infinite group is said to have **infinite order**.

The notation $|G|$ corresponds to other uses of the modulus sign to denote size.

Theorem 2.8 Lagrange's theorem

Let G be a finite group and H be a subgroup of G.
Then the order of H divides the order of G, i.e. $|H|$ divides $|G|$.

Proof

Since G has a finite number of elements, so does H.

Assume that the order of G is n and that the order of H is m.

Every element of G belongs to some coset, since any element a is equal to ae, which is an element of the coset aH.
Hence the distinct left cosets of H, being disjoint (by Theorem 2.7), form a partition of G.

Assume that there are l such disjoint cosets. Since each has m elements (by Exercise 2.9), the total number of elements in G is lm.
Therefore $n = lm$, which shows that m divides n. ∎

A set of subsets $S_1, S_2, \ldots, S_k$ of a set S forms a partition of S if their union is the whole of S and any two are disjoint.

3 NORMAL SUBGROUPS AND QUOTIENT GROUPS

3.1 Normal subgroups

In the last section we considered only *left* cosets of a subgroup. We might equally well have considered *right* cosets, in which case we would have obtained very similar results. For example, we would have obtained the following theorems.

> **Theorem 3.1**
>
> Let G be a group, H be a subgroup of G and a and b be elements of G. Then the right cosets Ha and Hb are equal if and only if $ab^{-1} \in H$.

Given a subgroup H of a group G and an element a of G, we define the *right coset* of H by a as the set
$$Ha = \{ha : h \in H\}.$$

The symmetry in a and b means that an equivalent condition is
$ba^{-1} \in H$.

> **Theorem 3.2**
>
> Let G be a group, H be a subgroup of G and a and b be elements of G. Then the right cosets Ha and Hb are either disjoint or equal.

The proofs of these results for right cosets are virtually identical to the ones already given for left cosets, and so we shall not write them out.

Given a particular subgroup of a group, it may or may not be the case that a left coset will be equal to the corresponding right coset.

Example 3.1

Let $G = S_3$, $H = \{e, (12)\}$. We saw in Section 2 that the left coset $(123)H$ is equal to $\{(123), (13)\}$.

On the other hand, the corresponding right coset, $H(123)$, is

$$\{(123), (12)(123)\} = \{(123), (23)\},$$

which is different from $(123)H$. ◆

However, it *is* true that there is a one–one correspondence *between* left cosets and right cosets.

Exercise 3.1

Let G be a group and H be a subgroup of G.
Show that ϕ, defined by

$$\phi(aH) = Ha^{-1}, \quad \text{for } a \in G,$$

is a *function*, from the set of left cosets of H to the set of right cosets of H, which is both *one-one* and *onto*.

Hint Since $\phi(aH)$ is defined in terms of a particular element a of the cosets aH, and not in terms of aH itself, to show that ϕ is *well-defined* (i.e. is indeed a function), you must show that if $aH = bH$ then $\phi(aH) = \phi(bH)$.

The solution to Exercise 3.1 shows that the number of left cosets of a subgroup is the same as the number of right cosets. Hence the following is a valid definition.

> **Definition 3.1 Index**
>
> Let H be a subgroup of a group G.
> If H has a finite number of left (or right) cosets in G, then this number of cosets is called the **index** of H in G.

Note that Lagrange's Theorem shows that the index of a subgroup H in a finite group G is the quotient

$$\frac{|G|}{|H|}.$$

Although the numbers of left and right cosets of a subgroup are the same, we have seen that, in general, corresponding left and right cosets are different. We now consider the special case where they *are* the same. We make the following definition.

Definition 3.2 Normal subgroup

Let H be a subgroup of a group G.
Then H is a **normal subgroup** of G if, for *every* $a \in G$, we have $aH = Ha$.

Note that we have *not* said that $ah = ha$ for all $a \in G$ and $h \in H$ *individually*, but just that the *sets* aH and Ha are the same.

Exercise 3.2

Prove that, in any group G, the subgroups $\{e\}$ and G are both normal subgroups.

Exercise 3.3

Show that $H = \{e, (123), (132)\}$ is a normal subgroup of the permutation group S_3.

Exercise 3.4

If H is a subgroup of index 2 of a group G, prove that H is a normal subgroup.

Using the definition to check if a subgroup is normal is not usually the most convenient way. There are several results of the form

    the subgroup H is normal if and only if H has some property

where the property is easier to check than the definition. One of these results is given in the following theorem.

Theorem 3.3

Let H be a subgroup of the group G.
Then H is a normal subgroup if and only if for each a in G we have $aHa^{-1} \subseteq H$.

$aHa^{-1} = \{aha^{-1} : h \in H\}$.

Proof

Only if

We are given that H is a normal subgroup of G and need to show that, for each element a of G, $aHa^{-1} \subseteq H$.

If x is in aHa^{-1}, then

$$x = ah_1a^{-1} \quad \text{for some } h_1 \in H.$$

But ah_1 is in aH, which is equal to Ha.
Therefore $ah_1 = h_2a$ for some h_2 in H. So

$$x = ah_1a^{-1} = h_2aa^{-1} = h_2 \in H.$$

Therefore $aHa^{-1} \subseteq H$.

If

This time we are given that $aHa^{-1} \subseteq H$ for each element a of G, and we need to show that $aH = Ha$.

As usual, to show that two sets are equal we show that each is a subset of the other.

We begin by showing that aH is a subset of Ha.

Let x be in aH, so that $x = ah_1$ for some h_1 in H.
The element $xa^{-1} = ah_1a^{-1}$ is in aHa^{-1}. Since aHa^{-1} is a subset of H, it follows that xa^{-1} is in H.
Therefore $xa^{-1} = h_2$ for some h_2 in H.
Hence $x = h_2a$, which is in Ha, and so aH is a subset of Ha. □

Exercise 3.5

Show that Ha is a subset of aH.

Proof of Theorem 3.3 continued

Therefore $aH = Ha$. ∎

Towards the end of Section 5, we shall be concerned with the normal subgroup *generated* by a given subset of a group, i.e. with the smallest normal subgroup containing the given set. The proof which shows that such a normal subgroup exists is virtually identical to the proof of Theorem 2.4.

The key step is to show that the intersection of normal subgroups is normal. (By Theorem 2.3, we already know that it is a subgroup.)

Exercise 3.6

Use Theorem 3.3 to show that the intersection of a collection of normal subgroups of a group G is again a normal subgroup of G.

Theorem 3.4

Let G be a group and S be a subset of (the underlying set of) G. Then there exists a unique smallest normal subgroup of G which contains S.

In other words, S generates a unique normal subgroup of G.

Proof

Consider the collection of all normal subgroups containing the set S. The collection is non-empty, since G is a normal subgroup containing S. Let N be the intersection of this collection.

By the solution to Exercise 3.6, N is a normal subgroup of G which, by the definition of intersection, contains S.

By the method of construction, N is the smallest such normal subgroup. By the definition of 'smallest', N is unique. ∎

Though Theorem 3.4 tells us that the required normal subgroup N exists, it does not tell us what the elements of N look like. As in the case of ordinary subgroups generated by a set S, there is a method of constructing N which rectifies this deficiency.

Let G be a group, S be a subset of G and N be the smallest normal subgroup of G containing S.

As before, we define
$$\hat{S} = S \cup \{e\} \cup \{s^{-1} : s \in S\}.$$

Since N is a subgroup containing S, it must contain $\hat{S}$. Therefore N must be the smallest normal subgroup of G containing $\hat{S}$.

Furthermore, since N is a *normal* subgroup of G containing $\hat{S}$, it follows that, for each a in G and s in $\hat{S}$, we must also have asa^{-1} in N. This prompts us to define $\tilde{S}$ as follows
$$\tilde{S} = \{asa^{-1} : s \in \hat{S},\ a \in G\}.$$

Exercise 3.7 _____

Prove that the inverse of every element of $\tilde{S}$ is also in $\tilde{S}$.

_____

Since N is a subgroup of G, the products of elements of $\tilde{S}$ must also be in N. Therefore we now consider the set of words of elements in $\tilde{S}$, denoted by $W(\tilde{S})$. In fact it turns out that $N = W(\tilde{S})$.

We already know, from the way it was constructed, that $W(\tilde{S})$ is contained in any normal subgroup which contains S. Therefore, in particular,
$$W(\tilde{S}) \subseteq N.$$

Hence, to show that $N = W(\tilde{S})$, we need to show that $N \subseteq W(\tilde{S})$.

In practice, what we do is prove that $W(\tilde{S})$ is a normal subgroup of G. The required result then follows by the minimality of N.

Theorem 3.5

Let G be a group and S be a subset of G. Define
$$\hat{S} = S \cup \{e\} \cup \{s^{-1} : s \in S\},$$
and
$$\tilde{S} = \{asa^{-1} : s \in \hat{S},\ a \in G\}.$$
Then the set of all words in elements of $\tilde{S}$,
$$W(\hat{S}) = \{s_1 s_2 \cdots s_n : n \in \mathbb{N},\ \text{where}\ s_i \in \hat{S},\ 1 \leq i \leq n\},$$
is the smallest normal subgroup of G which contains S.

Proof

From the comments preceding the statement of the theorem, we know that we need only verify that $W(\tilde{S})$ is a normal subgroup of G.

The set $\tilde{S}$ is non-empty since it contains $eee^{-1} = e$, the identity of G. The set is also closed under taking inverses (by Exercise 3.7). So, from Corollary 2.1, the set $W(\tilde{S})$, of all words in elements of $\tilde{S}$, is a subgroup of G.

By Theorem 3.3, to show that $W(\tilde{S})$ is a *normal* subgroup, we need only show that, for any elements w in $W(\tilde{S})$ and a in G, it follows that awa^{-1} is an element of $W(\tilde{S})$.

The key step in the proof uses the fact that
$$a(xy)a^{-1} = axa^{-1}aya^{-1}.$$
From the definition of $W(\tilde{S})$, we may take
$$w = (a_1 s_1 a_1^{-1})(a_2 s_2 a_2^{-1}) \cdots (a_n s_n a_n^{-1}), \quad n \in \mathbb{N},\ a_i \in G,\ s_i \in \hat{S},\ 1 \le i \le n.$$
Therefore
$$\begin{aligned}
awa^{-1} &= a(a_1 s_1 a_1^{-1})(a_2 s_2 a_2^{-1}) \cdots (a_n s_n a_n^{-1})a^{-1} \\
&= a(a_1 s_1 a_1^{-1})a^{-1} a(a_2 s_2 a_2^{-1})a^{-1} \cdots a(a_n s_n a_n^{-1})a^{-1} \\
&= ((aa_1)s_1(aa_1)^{-1})((aa_2)s_2(aa_2)^{-1}) \cdots ((aa_n)s_n(aa_n)^{-1}).
\end{aligned}$$

Now, since G is a group, the elements aa_i are also in G, and therefore awa^{-1} is an element of $W(\tilde{S})$. ∎

3.2 Quotient groups

From the previous section we know that the cosets of a subgroup H of a group G form a partition of G. When the subgroup concerned is a *normal* subgroup N, we can say more. We may define a binary operation on the set of cosets of N which makes the set of cosets into a group.

From here on in this course, when we refer to a coset of a subgroup we shall usually mean a left coset.

The set of cosets of the normal subgroup N in G is denoted by G/N, and the binary operation which makes this set of cosets into a group is defined as follows.

The symbol G/N is read either as 'the quotient of G by N' or, for reasons we shall explain shortly, as 'G mod N'.

Definition 3.3 Product of cosets

Let G be a group, N be a normal subgroup of G and a and b be elements of G. We define the **product** of the cosets aN and bN to be
$$(aN)(bN) = (ab)N.$$

Certainly, our definition ensures that the product of two cosets of N is a *coset* of N. But is this coset unique, i.e. is the product of cosets a *well-defined* operation? For example, as we have already seen, it is quite possible for $aN = a'N$ and $bN = b'N$ for some elements a' and b' different from a and b. If this is the case, the product $(aN)(bN) = (a'N)(b'N)$ has been defined to be both $(ab)N$ and $(a'b')N$. So our definition will only produce a well-defined binary operation if each such product is uniquely defined, i.e. if $(ab)N = (a'b')N$.

The justification that the product of two cosets is unique is provided by the following theorem.

Theorem 3.6

Let G be a group, N be a normal subgroup of G and a, a', b and b' be elements of G such that
$$aN = a'N \quad \text{and} \quad bN = b'N.$$
Then
$$(ab)N = (a'b')N.$$

Proof

By Theorem 2.6, to show that $(ab)N = (a'b')N$ it is enough to show that $(ab)^{-1}(a'b')$ is in N.

Now,
$$(ab)^{-1}a'b' = b^{-1}a^{-1}a'b'. \tag{3.1}$$

Since $aN = a'N$, Theorem 2.6 tells us that
$$a^{-1}a' \in N,$$
and therefore
$$b^{-1}a^{-1}a' \in b^{-1}N.$$

But $b^{-1}N = Nb^{-1}$, as N is normal, and therefore
$$b^{-1}a^{-1}a' = nb^{-1}, \quad \text{for some } n \text{ in } N.$$

So, from Equation 3.1,
$$(ab)^{-1}a'b' = nb^{-1}b', \quad \text{for some } n \text{ in } N. \tag{3.2}$$

But Theorem 2.6 also tells us that
$$b^{-1}b' \in N,$$
and so Equation 3.2 tells us that $(ab)^{-1}a'b'$ is a product of two elements, n and $b^{-1}b'$, of N. Since N is a group, it follows that
$$(ab)^{-1}a'b' \in N. \qquad \blacksquare$$

Theorem 3.6 is the first step in proving that the set of cosets G/N, with binary operation defined by $(aN)(bN) = (ab)N$, is a group.

All we now need to do is to check that the group axioms are satisfied.

The closure axiom is clearly satisfied, since, as we remarked above, $(ab)N \in G/N$.

> In future we shall often just write $aNbN$ instead of the more formal $(aN)(bN)$, and abN instead of $(ab)N$.

Exercise 3.8

Check that the binary operation that we have defined on the set of cosets G/N is associative.

Exercise 3.9

Prove that the coset N ($= eN$) is the identity of G/N.

Exercise 3.10

Prove that the inverse of the coset aN of G/N is $a^{-1}N$.

The results of the previous exercises enable us to make the following definition.

Definition 3.4 Quotient group

Let N be a normal subgroup of the group G.
The **quotient group of G with respect to N** is defined as follows.

Set The set of all cosets of N, i.e. the set G/N.

Operation If aN and bN are elements of G/N then
$$(aN)(bN) = (ab)N.$$

Exercise 3.11

Let G be the group of symmetries of the square, denoted by D_4. In other words,
$$G = \{r^m s^n : m = 0, 1, 2, 3, \ n = 0, 1; \ r^4 = s^2 = e, \ sr = r^3 s\}$$
$$= \{e, r, r^2, r^3, s, rs, r^2 s, r^3 s\}.$$

Let N be the set $\{e, r^2\}$.

(a) Prove that N is a normal subgroup of G.

(b) Find the set of left cosets of N in G.

(c) Using the notation given in the solution to part (b), write out the Cayley table of the quotient group G/N.

The elements of D_4 may be generated by a rotation r through $\pi/2$ about the centre of the square and a reflection s in an axis through the centre of the square and parallel to two of its sides.

You might have noticed that, since aN and bN are both *subsets* of G, we could have defined their product $(aN)(bN)$ as the product of subsets, i.e.
$$(aN)(bN) = \{xy : x \in aN, \ y \in bN\}.$$

This definition certainly has the advantage that this product is well-defined, since it depends on the cosets as a whole rather than on the particular elements a and b in the cosets. On the other hand, the difficulty with this definition is that we have no idea whether or not the product defined is a coset of N. In fact, the two definitions *do* produce the same product.

If you would like to confirm this, you should try the following exercise.

Exercise 3.12

Show that both definitions of $(aN)(bN)$ produce the same result. In other words, prove that if N is a normal subgroup of a group G then
$$\{xy : x \in aN, \ y \in bN\} = \{(ab)n : n \in N\}.$$

To illustrate the quotient group construction, we consider the subgroup
$$N = 4\mathbb{Z} = \{4z : z \in \mathbb{Z}\}$$
of the group $(\mathbb{Z}, +)$.

First, N is a subgroup of $\mathbb{Z}$. The proof is straightforward and is virtually the same as as for $2\mathbb{Z}$, which we discussed in Section 2.

Next, N is normal. If $z \in \mathbb{Z}$, the normality check $aHa^{-1} \subseteq H$ from Theorem 3.3, in additive notation, and using z and N in place of a and H, becomes
$$z + N + (-z) \subseteq N.$$

We show this holds as follows. Any element of $z + N + (-z)$ is of the form
$$z + n + (-z), \quad \text{for some } n \text{ in } N.$$

However, $\mathbb{Z}$ is Abelian, and so
$$z + n + (-z) = z + (-z) + n = n \in N,$$
and therefore N is normal.

A group G with binary operation $\circ$ is Abelian if $x \circ y = y \circ x$ for all pairs x, y of elements in G.

Now, let us determine the distinct cosets of N, i.e. the elements of $\mathbb{Z}/N = \mathbb{Z}/4\mathbb{Z}$.

By Theorem 2.6, the cosets $a + N$ and $b + N$ are equal if and only if $a + (-b) \in N$; in other words if and only if a and b differ by a multiple of 4. Thus elements of $\mathbb{Z}$ are in the same coset if and only if they are equal mod 4.

In additive notation, the distinct cosets are therefore:

$$4\mathbb{Z} = 0 + 4\mathbb{Z} = \{4z : z \in \mathbb{Z}\};$$
$$1 + 4\mathbb{Z} = \{1 + 4z : z \in \mathbb{Z}\};$$
$$2 + 4\mathbb{Z} = \{2 + 4z : z \in \mathbb{Z}\};$$
$$3 + 4\mathbb{Z} = \{3 + 4z : z \in \mathbb{Z}\}.$$

It is clear that addition of the cosets $a + N$ and $b + N$ corresponds to addition of a and b mod 4.

The language here is often carried over to the general case. The quotient group G/N is often referred to as 'G mod N'. Calculations in the quotient group G/N are like those in G, except that elements of N are treated as if they were the identity. Thus the quotient group is, in some sense, the group we obtain starting from the group G but also insisting that, for each element $n \in N$, the relation $n = e$ is satisfied. We shall say more about this aspect of quotient groups later in the unit.

4 ISOMORPHISMS AND HOMOMORPHISMS

4.1 Isomorphisms

In *Unit IB2*, there were several examples of groups each having four elements.

Firstly, in the Introduction, you met the Klein group, V, an abstract group with Cayley table

∘	e	a	b	c
e	e	a	b	c
a	a	e	c	b
b	b	c	e	a
c	c	b	a	e

Secondly, in Section 1, you met the group of symmetries of the rectangle, $\Gamma(\square)$, with Cayley table

∘	e	r	h	v
e	e	r	h	v
r	r	e	v	h
h	h	v	e	r
v	v	h	r	e

In Section 1 of *Unit IB2*, you also saw that $\Gamma(\square)$ could also be represented as a group of orthogonal matrices and as a group of permutations.

Thirdly, in Section 4, you met the group $\{1, 2, 3, 4\}$ of non-zero integers with binary operation multiplication modulo 5, $(\mathbb{Z}_5^*, \times_5)$, with Cayley table

$\times_5$	1	2	3	4
1	1	2	3	4
2	2	4	1	3
3	3	1	4	2
4	4	3	2	1

From now on we shall use $\times_n$ to represent multiplication modulo n.

In Section 4 of *Unit IB2*, you also met the groups $\{0, 1, \cdots, n-1\}$ of integers with binary operation addition modulo n, $(\mathbb{Z}_n, +_n)$, for all positive integers n.

From now on we shall use $+_n$ to represent addition modulo n.

Exercise 4.1

Write out the Cayley table for the group $(\mathbb{Z}_4, +_4)$.

Exercise 4.2

Prove that the subgroup R of the dihedral group D_4 generated by the rotation r has four elements, and write down its Cayley table.

Remember that D_4 is the group of symmetries of the square.

The main question we shall tackle in this subsection is whether these four-element groups (and other collections of groups with the same number of elements) are 'essentially the same' group. The first problem to be solved in answering this question is to define formally what we mean by saying that groups are 'essentially the same'. We shall define two groups to be 'essentially the same' if they are *isomorphic*, a concept that you have met informally in *Units IB2* and *IB3* and, perhaps, formally in your previous mathematical studies. However, we shall delay the formal definition until we have looked at our examples in a little more detail.

Clearly the five groups above, namely $(V, \circ), (\Gamma(\square), \circ), (\mathbb{Z}_5^*, \times_5), (\mathbb{Z}_4, +_4)$ and $(R, \circ)$, are not the same as one another, in the sense of being identical, since the underlying sets are different and each has a different binary operation. However, as we noted in *Unit IB2*, the first two — $(V, \circ)$ and $(\Gamma(\square), \circ)$ — are essentially the same, in that their Cayley tables have the same pattern: if in the Cayley table for $(V, \circ)$ we replace a by r, b by h and c by v, then the first table becomes the second. Similarly, the Cayley tables of the fourth and fifth groups have the same pattern as one another:

$+_4$	0	1	2	3
0	0	1	2	3
1	1	2	3	0
2	2	3	0	1
3	3	0	1	2

$(\mathbb{Z}_4, +_4)$

$\circ$	e	r	r^2	r^3
e	e	r	r^2	r^3
r	r	r^2	r^3	e
r^2	r^2	r^3	e	r
r^3	r^3	e	r	r^2

$(R, \circ)$

If, in the Cayley table for $(\mathbb{Z}_4, +_4)$, we replace $+_4$ by $\circ$, 0 by e, 1 by r, 2 by r^2 and 3 by r^3, then the first table becomes the second.

Now the above replacements in the headings of the Cayley tables define two one–one, onto functions:

$$\psi : V = \{e, a, b, c\} \to \Gamma(\square) = \{e, r, h, v\};$$
$$\phi : \mathbb{Z}_4 = \{0, 1, 2, 3\} \to R = \{e, r, r^2, r^3\}.$$

The fact that these replacements applied to the body of the first table in each case produces the second means that these replacements preserve the effect of the binary operations. In other words, the image of the combination of elements in the first table in each case is the same as the combination of their images in the second table. That is, for example, in the second case, for all x and y in $\mathbb{Z}_4$, we have

$$\phi(x +_4 y) = \phi(x) \circ \phi(y).$$

This is what we shall mean by groups being 'essentially the same', and if this is the case we shall say that they are isomorphic.

Definition 4.1 Isomorphism

The groups $(G, \circ)$ and $(H, *)$ are **isomorphic** if and only if there exists a one–one, onto function $\phi : G \to H$ such that, for all a and b in G, we have

$$\phi(a \circ b) = \phi(a) * \phi(b).$$

Such a function is called an **isomorphism** from G to H.

Using multiplicative notation for both groups, the last statement would be written as

$$\phi(ab) = \phi(a)\phi(b).$$

This is the notation which we shall most commonly use. Where the binary operation in each group is addition, the corresponding statement is

$$\phi(a+b) = \phi(a) + \phi(b).$$

Exercise 4.3

Prove that the function

$$\phi : \mathbb{Z} \to 2\mathbb{Z}$$
$$n \mapsto 2n$$

defines an isomorphism from $(\mathbb{Z}, +)$ to $(2\mathbb{Z}, +)$.

We proved that $(2\mathbb{Z}, +)$ was a group in Section 2.

You may have noticed that, in the above definition, we moved from the statement 'G and H are isomorphic', to the statement 'ϕ is an isomorphism from G to H'. The first statement is symmetrical in G and H whereas the second is not. Fortunately there is no conflict here, as the following theorem shows.

Theorem 4.1

Let G and H be groups and let ϕ be an isomorphism from G to H. Then ϕ^{-1} is an isomorphism from H to G.

Proof

Since the function ϕ is both one–one and onto, the function ϕ^{-1} exists and is a one–one and onto function from H to G.

We need to prove that, for any two elements x and y of the group H, we have

$$\phi^{-1}(xy) = \phi^{-1}(x)\phi^{-1}(y).$$

Since the function ϕ is onto, there exist elements a and b in G such that $\phi(a) = x$ and $\phi(b) = y$ or, to put it another way, $a = \phi^{-1}(x)$ and $b = \phi^{-1}(y)$.

Also, since ϕ is an isomorphism, we have $\phi(ab) = \phi(a)\phi(b)$.

Therefore

$$\begin{aligned}\phi^{-1}(xy) &= \phi^{-1}\left(\phi(a)\phi(b)\right) \\ &= \phi^{-1}\left(\phi(ab)\right) \\ &= ab \\ &= \phi^{-1}(x)\phi^{-1}(y).\end{aligned}$$ ∎

Returning to our five example groups, we have shown so far that $(V, \circ)$ and $(\Gamma(\square), \circ)$ are isomorphic and that $(\mathbb{Z}_4, +_4)$ and $(R, \circ)$ are isomorphic. Are there any other isomorphisms among the five groups?

We have already seen, in Section 4 of *Unit IB2*, that $(\mathbb{Z}_5^*, \times_5)$ and $(\mathbb{Z}_4, +_4)$ are both cyclic groups with four elements, so presumably they too are isomorphic. When we look at their Cayley tables, however, they do not seem to have the same pattern as one another:

$+_4$	0	1	2	3
0	0	1	2	3
1	1	2	3	0
2	2	3	0	1
3	3	0	1	2

$\times_5$	1	2	3	4
1	1	2	3	4
2	2	4	1	3
3	3	1	4	2
4	4	3	2	1

$(\mathbb{Z}_4, +_4)$ $(\mathbb{Z}_5^*, \times_5)$

However, this might just mean that the order of the elements in the headings of tables means that we are considering the wrong function ϕ. So far we have no clue as to how to look for suitable functions ϕ. On the other hand, have said that two isomorphic groups are 'essentially the same', so it should follow that corresponding elements will have the same properties. In particular, the identity of one group should correspond to the identity of the other. This is in fact the case.

> **Theorem 4.2**
>
> Let G and H be groups, and let ϕ be an isomorphism from G to H. Then, if e is the identity of G, $\phi(e)$ is the identity of H.

Proof

As e is the identity for the group G, we know that $ea = a$ for any element a in G.

Since ϕ is a isomorphism, this gives $\phi(e)\phi(a) = \phi(a)$.

Therefore $\phi(e)$ behaves like a left identity for the element $\phi(a)$ of H, and so (by Theorem 1.7) is the identity of H. ∎

The expected result also holds for inverses.

Exercise 4.4

Let G and H be groups, and let ϕ be an isomorphism from G to H. If the element a of G has inverse a^{-1}, show that $\phi(a)$ has inverse $\phi(a^{-1})$. In other words, show that

$$(\phi(a))^{-1} = \phi(a^{-1}).$$

Hint As in the proof of Theorem 4.2, use one of the results from Section 1.

So now at least we know that, if there is an isomorphism ϕ from $(\mathbb{Z}_4, +_4)$ to $(\mathbb{Z}_5^*, \times_5)$, then the identity, 0, in $\mathbb{Z}_4$ must correspond to the identity, 1, in $\mathbb{Z}_5^*$ and that inverses must correspond to inverses.

In *Unit IB2* we showed that $(\mathbb{Z}_5^*, \times_5)$ is a cyclic group by verifying that the element 2 is a generator. If there is an isomorphism from $(\mathbb{Z}_4, +_4)$ to this group, then surely a generator in one group must correspond to a generator in the other. So we might try letting the generator 1 in $\mathbb{Z}_4$ correspond to the generator 2 in $\mathbb{Z}_5^*$. Since 3 is the inverse of 1 in $\mathbb{Z}_4$ and 3 is the inverse of 2 in $\mathbb{Z}_5^*$, the 3s should also correspond. This leaves the 2 in $\mathbb{Z}_4$ to correspond to the 4 in $\mathbb{Z}_5^*$.

In fact, when we write the Cayley table of $(\mathbb{Z}_5^*, \times_5)$ as

$\times_5$	1	2	4	3
1	1	2	4	3
2	2	4	3	1
4	4	3	1	2
3	3	1	2	4

we find that it does have the same pattern as the Cayley table for $(\mathbb{Z}_4, +_4)$.

So the function $\phi: \mathbb{Z}_4 \to \mathbb{Z}_5^*$ defined by

$\phi(0) = 1$
$\phi(1) = 2$
$\phi(2) = 4$
$\phi(3) = 3$

is an isomorphism from $(\mathbb{Z}_4, +_4)$ to $(\mathbb{Z}_5^*, \times_5)$.

Exercise 4.5

Find another isomorphism from $(\mathbb{Z}_4, +_4)$ to $(\mathbb{Z}_5^*, \times_5)$.

We have seen that $(\mathbb{Z}_4, +_4)$ is isomorphic to $(R, \circ)$ and to $(\mathbb{Z}_5^*, \times_5)$, and we can deduce therefore that $(R, \circ)$ and $(\mathbb{Z}_5^*, \times_5)$ are isomorphic. We have also seen that $(V, \circ)$ and $(\Gamma(\square), \circ)$ are isomorphic. But is there an isomorphism between $(\mathbb{Z}_4, +_4)$ and $(\Gamma(\square), \circ)$, say, in which case we could deduce that all five groups are isomorphic? The answer is no, but how do we go about showing this?

This follows because the inverse of an isomorphism is an isomorphism, as is the composite of two isomorphisms.

In order to determine whether or not $(\Gamma(\square), \circ)$ is isomorphic to $(\mathbb{Z}_4, +_4)$, it seems that we would need to consider all possible ways of writing out its Cayley table and see whether any of the rearrangements has the same pattern as our table for $(\mathbb{Z}_4, +_4)$. (Equivalently, we would have to consider all one–one, onto functions from $\mathbb{Z}_4$ to $\Gamma(\square)$ and check whether any of these satisfies the isomorphism property.)

At first glance this seems to mean checking $4! = 24$ tables, since that is the number of ways of rearranging four elements. We already know that identity elements must correspond and so must inverses, which cuts down the work somewhat. Nevertheless it would still seem to involve a considerable amount of effort.

Clearly this approach is not the way to prove that large finite groups are, or are not, isomorphic. For infinite groups this approach is impossible, since there are an infinite number of possible ϕs to consider.

In practice it is often easier to prove that groups are *not* isomorphic than that they *are*. But why, you might well ask, might we want to know that two groups are not isomorphic?

Well, one of the main objectives of this course is to classify certain classes of objects. In the Geometry stream, we classify tilings, friezes and wallpaper patterns and, in the Groups stream, we classify Abelian groups having a finite number of generators. And, to classify a set of objects, we have to carry out three tasks.

Firstly, we divide the set of objects into a number of distinct subsets such that two objects are in the same subset if they are 'essentially the same' in an appropriate sense for the objects being considered. For friezes, for example, it means having the same frieze group. For groups it means being isomorphic, so that, for example, all cyclic groups of order four would form a subset.

We *partition* the set of objects.

Secondly, we give a particular representative for each subset. In other words, we give what we regard as a typical object of its type. For friezes of Type 6, for example, we might choose a plain rectangular frieze as being representative. For the subset of cyclic groups of order four, we might, for example, give the representative $(\mathbb{Z}_4, +_4)$.

Thirdly, we provide an algorithm which enables us to decide in which subset a given object lies. An example of such an algorithm, for classifying friezes, was given in *Unit IB3*. An example of such an algorithm for Abelian groups having a finite number of generators will be given later in the course.

In order to perform such a classification for groups, it is important to know not only when two groups are isomorphic, that is when they belong to the same subset, but also when they are not isomorphic, and therefore belong to different subsets.

Since isomorphic groups are 'essentially the same', they, and their corresponding elements, must have identical algebraic properties. So, one way that we can show that two groups are not isomorphic is to show that the groups have different algebraic properties.

We use the word 'algebraic' here since it is possible for two groups to be isomorphic and yet to have different *geometric* properties, as we saw in the case of the frieze groups in *Unit IB3*.

One property of a finite group is the number of elements that is has. Thus one way in which finite groups might be different is when they have different numbers of elements. In this case it is not possible to construct a one–one, onto function from one group to the other, and so they cannot be isomorphic. Similarly, it is not possible for a finite and an infinite group to be isomorphic.

However, when groups have the same number of elements, or are both infinite, we need to consider other properties, that is other than the number of elements, to show that they are not isomorphic.

Exercise 4.6

Let G and H be groups, and let ϕ be an isomorphism from G to H.

(a) Prove that if G is Abelian, then so is H.

(b) Prove that the groups $\mathbb{Z}_6$ and S_3 are not isomorphic.

Remember that a group G is Abelian if $ab = ba$ for all pairs a, b of elements in G.

The solution to Exercise 4.6 provides a first consequence of two groups being isomorphic: if one of them is Abelian, then so is the other. There is the corresponding negative result: an Abelian group and a non-Abelian group cannot be isomorphic. This result was used in Section 4 of *Unit IB3* to prove that certain pairs of frieze groups are not isomorphic.

Another useful property in deciding whether or not groups are isomorphic (which was also used in Section 4 of *Unit IB3*) is the number of elements in the group having a given order. The formal definition of the order of an element is as follows.

Definition 4.2 Order of group element

Let a be an element of a group G having identity e.
Then a has **order** n if and only if n is the smallest positive integer such that $a^n = e$.
If no such positive integer exists, then the element a is said to have **infinite order**.

We use the notation $|a|$ to denote the order of an element a of finite order.

This definition is equivalent to the slightly different one given in *Unit IB2*.

We have now used the word 'order' in two senses: in Section 2 we defined the order of a group and above we have defined the order of a group element. These two are consistent in that the above definition says that the order of an element is the order of the cyclic subgroup that it generates. Even the use of the modulus notation is shared.

Recall that the definition of the order of a group element in *Unit IB2* was couched in terms of the order of the cyclic subgroup that it generates.

Theorem 4.3

Let G and H be groups and let ϕ be an isomorphism from G to H. Then, if the element a of G has finite order n, so does the element $\phi(a)$ of H.

Proof

Let e be the identity of G, f be the identity of H and m be the order of $\phi(a)$.

Since a has order n, we know that $a^n = e$ and so $\phi(a^n) = \phi(e)$. Now, using Theorem 4.2, we can say that

$$(\phi(a))^n = \phi(a^n) = \phi(e) = f.$$

Here we are using a straightforward extension of the property that
$$\phi(ab) = \phi(a)\phi(b)$$
to
$$\phi(a^n) = (\phi(a))^n.$$

Since the order of $\phi(a)$ is m, this last result tells us that $m \leq n$, since the order of $\phi(a)$ is the *smallest* positive integer power of $\phi(a)$ which produces the identity f.

As $\phi(a)$ has order m, we know that $(\phi(a))^m = f$, so

$$\phi(a^m) = (\phi(a))^m = f = \phi(e).$$

However, ϕ is a one–one function, so this implies that $a^m = e$.

But a has order n and so, by the minimality of n, $n \leq m$. Combining this with the fact that $m \leq n$ gives $m = n$, and so the order of a is the same as the order of $\phi(a)$. ∎

Exercise 4.7

Let G and H be groups and let ϕ be an isomorphism from G to H.
Prove that, if the element a of G has infinite order, then so does $\phi(a)$.

Hint Use the fact that ϕ^{-1} is an isomorphism from H to G to show that, if $\phi(a)$ has finite order, then so does a.

Finally, Theorem 4.3 enables us to say that the groups $\mathbb{Z}_4$ and $\Gamma(\square)$ are not isomorphic. The group $\mathbb{Z}_4$ has two elements of order four (1 and 3), whereas $\Gamma(\square)$ has no elements of order four (each of the non-identity elements has order two). So having two elements of order four is a property which one group has and the other does not.

Perhaps you think that this result would have been easier to obtain by remarking that one of the groups is cyclic and the other one is not. In essence it is the last theorem which justifies that this is a correct argument. One group has the property of having an element of order four whereas the other does not. We shall see, however, that this theorem has a wider application than just to cyclic groups.

Exercise 4.8

Prove that the groups S_3 and $\mathbb{Z}_2 \times \mathbb{Z}_3$ are not isomorphic.

4.2 Homomorphisms

For two groups $(G, \circ)$ and $(H, *)$ to be isomorphic there must exist a one–one and onto function $\phi: G \to H$ such that, for all a and b in G, we have $\phi(a \circ b) = \phi(a) * \phi(b)$.

If we abandon the conditions that ϕ must be one–one and onto, then ϕ is called a *homomorphism* from $(G, \circ)$ to $(H, *)$.

Definition 4.3 *Homomorphism*

Let $(G, \circ)$ and $(H, *)$ be groups. A function $\phi: G \to H$ is a **homomorphism** from G to H if, for all a and b in G, we have

$$\phi(a \circ b) = \phi(a) * \phi(b).$$

The property
$$\phi(a \circ b) = \phi(a) * \phi(b).$$
is referred to as the **morphism** or **homomorphism property**.

So isomorphisms are also homomorphisms, but not all homomorphisms need be isomorphisms, since the function may fail to be one–one or may fail to be onto (or indeed may fail to be either one–one or onto).

Example 4.1

Consider the groups $(\mathbb{Z}, +)$ and $(\mathbb{Z}_4, +_4)$ and the function ϕ from $\mathbb{Z}$ to $\mathbb{Z}_4$ defined by:

$\phi(n) = 0$ if n is even;
$\phi(n) = 2$ if n is odd.

Clearly, ϕ is a function which is neither one–one (since, for example, $\phi(0) = \phi(2) = 0$) nor onto (since, for example, 1 is not in the image set of ϕ).

To check that ϕ is a homomorphism we must check the morphism property, i.e. that, for all m and n in $\mathbb{Z}$,

$$\phi(m+n) = \phi(m) +_4 \phi(n).$$

Since ϕ was defined by two cases, this can be done by considering just four cases, determined by whether m and n are even or odd.

If m and n are both even, so is $m+n$ and therefore

$$\phi(m+n) = 0 = 0 +_4 0 = \phi(m) +_4 \phi(n).$$

If m is odd and n is even, then $m+n$ is odd and therefore

$$\phi(m+n) = 2 = 2 +_4 0 = \phi(m) +_4 \phi(n).$$

If m is even and n is odd, then $m+n$ is odd and therefore

$$\phi(m+n) = 2 = 0 +_4 2 = \phi(m) +_4 \phi(n).$$

If m and n are both odd, then $m+n$ is even and therefore

$$\phi(m+n) = 0 = 2 +_4 2 = \phi(m) +_4 \phi(n).$$

So for all m and n in $\mathbb{Z}$, we have that

$$\phi(m+n) = \phi(m) +_4 \phi(n).$$

This is an example of a homomorphism which is neither one–one nor onto. ♦

As you may recall from your previous mathematical studies, the set of elements that are mapped to the identity in the codomain is called the *kernel* of the homomorphism.

Definition 4.4 Kernel

Let $\phi: G \to H$ be a homomorphism and f be the identity of H.
Then the **kernel** of ϕ, denoted by $\text{Ker}(\phi)$, is defined by

$\text{Ker}(\phi) = \{x \in G : \phi(x) = f\}$.

In Example 4.1, the kernel consists of all those elements of $\mathbb{Z}$ mapped to 0, the identity of $\mathbb{Z}_4$. These are the even integers; that is,

$$\text{Ker}(\phi) = 2\mathbb{Z}.$$

Note that this is a subgroup of the domain.

Definition 4.5 Image

Let $\phi: G \to H$ be a homomorphism.
Then the **image** of ϕ, denoted by $\text{Im}(\phi)$, is defined by

$\text{Im}(\phi) = \{h \in H : h = \phi(x) \text{ for some } x \in G\}$.

In the case of Example 4.1,

$$\text{Im}(\phi) = \{0, 2\},$$

which is a subgroup of the codomain.

Exercise 4.9

This exercise concerns the following function:

$$\phi : \mathbb{Z} \to \mathbb{Z}$$
$$n \mapsto 3n$$

(a) Show that ϕ is a homomorphism from $(\mathbb{Z}, +)$ to $(\mathbb{Z}, +)$, but is not an isomorphism.

(b) Find the kernel of ϕ and show that it is a subgroup of the domain.

(c) Find the image of ϕ and show that it is a subgroup of the codomain.

The results of the previous exercise generalize, and we can say rather more about the kernel than just that it is a subgroup of the domain.

Theorem 4.4

Let $\phi : G \to H$ be a homomorphism from G to H. Then:

(a) the kernel of ϕ is a normal subgroup of G;

(b) the image of ϕ is a subgroup of H.

Proof

We use multiplicative notation throughout.

Let K and I be the kernel and image of ϕ respectively, and let e and f be the identities of G and H respectively.

(a) *Closure*

If k and l are in K, then $\phi(k) = \phi(l) = f$. Therefore

$$\phi(kl) = \phi(k)\phi(l) = ff = f,$$

and so $kl \in K$.

Identity

We have already shown in Theorem 4.2 that, if e is the identity of G, then $\phi(e)$ is the identity f of H for any *isomorphism* ϕ from G to H. But the proof of Theorem 4.2 made use only of the morphism part of the definition of an isomorphism, and so holds for homomorphisms too. Thus $e \in K$.

Inverses

We know from Exercise 4.4 that, for all $x \in G$, we have

$$\phi(x^{-1}) = (\phi(x))^{-1},$$

for any *isomorphism* ϕ from G to H. But the solution to Exercise 4.4 made use only of the morphism part of the definition of an isomorphism, and so holds for homomorphisms too. Hence, if $k \in K$ then $\phi(k) = f$, so

$$\phi(k^{-1}) = f^{-1} = f.$$

Thus $k^{-1} \in K$.

Hence the kernel, K, is a subgroup of the domain, G.

Normality

If $a \in G$ and $k \in K$, then

$$\phi\left(aka^{-1}\right) = \phi(a)\phi(k)\phi(a^{-1})$$
$$= \phi(a) f \left(\phi(a)\right)^{-1}$$
$$= \phi(a) \left(\phi(a)\right)^{-1}$$
$$= f.$$

Therefore, for all a in G, $aKa^{-1} \subseteq K$. So K is a normal subgroup of G.

(b) *Closure*

Let i and j be in I. Therefore there exist a and b in G such that $\phi(a) = i$ and $\phi(b) = j$. Now ab is in G and

$$\phi(ab) = \phi(a)\phi(b) = ij,$$

showing that $ij \in I$.

Identity

Since $\phi(e) = f$, by the homomorphism version of Theorem 4.2 it follows that $f \in I$.

Inverses

If $i \in I$, then $i = \phi(x)$ for some $x \in G$. Now, by the homomorphism version of Exercise 4.4,

$$\phi(x^{-1}) = (\phi(x))^{-1} = i^{-1}.$$

Therefore, since $x^{-1} \in G$, $i^{-1} \in I$. ∎

Theorem 4.4 tells us that a homomorphism ϕ from G to H gives rise to a normal subgroup $K = \text{Ker}(\phi)$ of the domain G. This normal subgroup can be used to define the quotient group G/K.

We also have the subgroup $I = \text{Im}(\phi)$ of the codomain H.

In fact, the quotient G/K and the image I are isomorphic. Before giving a general proof, we illustrate this result by expanding an example considered earlier.

Example 4.2

The example is the homomorphism ϕ from $\mathbb{Z}$ to $\mathbb{Z}_4$ defined in Example 4.1 by:

$\phi(n) = 0$ if n is even;
$\phi(n) = 2$ if n is odd.

We have already found that, for this example:

$K = 2\mathbb{Z}$;
$I = \{0, 2\}$.

The cosets of $K = 2\mathbb{Z}$ are:

$0 + K = 2\mathbb{Z}$,   the set of even integers;
$1 + K = 1 + 2\mathbb{Z}$,   the set of odd integers.

Since both $\mathbb{Z}/K$ and I are cyclic groups of order 2, they must be isomorphic. Because identities map to identities under isomorphisms, the only possible isomorphism ψ is given by:

$\psi(0 + K) = 0 = \phi(0)$;
$\psi(1 + K) = 2 = \phi(1)$.

Thus

$$\psi(x + K) = \phi(x). \qquad \blacklozenge$$

The last observation in the above example suggests the definition of the isomorphism

$$\psi : G/K \to I$$

in the general case.

> **Theorem 4.5 First isomorphism theorem**
>
> Let $\phi : G \to H$ be a homomorphism from the group G to the group H, with kernel K and image I. Then ψ defined by
>
> $$\psi : G/K \to I$$
> $$xK \mapsto \phi(x)$$
>
> is an isomorphism.

We shall prove other isomorphism theorems later in the Groups stream.

Proof

Firstly, since ψ is defined in terms of a representative x of the coset xK, not in terms of the coset itself, we must check that ψ is *well-defined* (i.e. is a *function*).

We need to show that, for any two elements a and b in G, if $aK = bK$ then $\psi(aK) = \psi(bK)$, that is $\phi(a) = \phi(b)$.

Suppose $aK = bK$. Then, by Theorem 2.6,

$$a^{-1}b \in K$$

and so, by the definition of K, $\phi(a^{-1}b) = f$, where f is the identity of H. So

$$\phi(a^{-1})\phi(b) = (\phi(a))^{-1}\phi(b) = f.$$

We use the homomorphism version of Exercise 4.4.

Hence, multiplying both sides on the left by $\phi(a)$, we obtain

$$\phi(a) = \phi(b),$$

as required.

Next, we must show that ψ is *one–one*.

Suppose that $\psi(aK) = \psi(bK)$, that is $\phi(a) = \phi(b)$. We must show that $aK = bK$.

Since $\phi(a) = \phi(b)$, reversing the above argument gives

$$\phi(a^{-1}b) = f.$$

This means that

$$a^{-1}b \in K$$

and so, by Theorem 2.6, $aK = bK$.

Next, we must show that ψ is *onto*.

Any element i of I is of the form $\phi(a)$ for some element a of G. Since $\psi(aK) = \phi(a) = i$, it follows that ψ is onto.

Lastly, we must show that ψ has the *morphism property*. □

Exercise 4.10

Show that ψ has the morphism property.

Proof of Theorem 4.5 continued

Hence the theorem is proved. ∎

Theorem 4.4 showed that all kernels are normal subgroups. Our final result in this section shows that the converse is also true: every normal subgroup is the kernel of a suitably defined homomorphism. Thus a subgroup is normal if and only if it is the kernel of a homomorphism.

If N is a normal subgroup of a group G, then we can form the quotient group G/N. This quotient will be the codomain for the homomorphism of which N is the kernel. There is a 'natural' association of elements of G with cosets of N: each element $a \in G$ belongs to the coset aN.

When we formalize these ideas, we obtain the following theorem.

Theorem 4.6

Let N be a normal subgroup of a group G.
Then ϕ defined by $\phi(a) = aN$ is a homomorphism from G onto the quotient group G/N.
Furthermore, the kernel of ϕ is N.

Proof

By definition, ϕ is certainly a function from G to G/N.

Next we check the morphism property. For any a and b in G, we have

$$\phi(ab) = abN = aNbN = \phi(a)\phi(b).$$

Thus ϕ is a homomorphism from G to G/N.

We now show that ϕ is onto G/N. Any element of G/N is a coset aN for some element a of G. As such it is the image of the element a under ϕ, because we defined $\phi(a) = aN$. Hence the function ϕ is onto.

Finally, we find the kernel of ϕ. The identity element of G/N is N. Therefore a is an element of the kernel of ϕ if and only if $\phi(a) = N$. However, $\phi(a) = aN$, and (by Lemma 2.1) $aN = N$ if and only if $a \in N$. Hence

$$\text{Ker}(\phi) = N. \qquad \blacksquare$$

Definition 4.6 Natural homomorphism

If N is a normal subgroup of a group G, then the homomorphism defined by

$$\phi : G \to G/N$$
$$a \mapsto aN$$

is the **natural homomorphism** from G onto G/N.

5 GENERATORS AND RELATIONS (AUDIO-TAPE SECTION)

In *Units IB2* and *IB3* we described certain groups in terms of elements in 'standard form' and defining relations between elements. For example, we described D_6 as

$$D_6 = \{r^m s^n : m = 0, \ldots, 5,\ n = 0, 1;\ r^6 = s^2 = e,\ sr = r^5 s\}.$$

In Section 2 of this unit we described groups in terms of the elements that generate the group. In the notation of that section, we can write D_6 as

$$D_6 = \langle r, s \rangle.$$

We can combine these two ideas to obtain a description of groups in terms of generators and relations. In the case of D_6 we obtain

$$D_6 = \langle r, s : r^6 = s^2 = e,\ sr = r^5 s \rangle.$$

We could now go on to give descriptions of all the example groups we have seen in terms of generators and relations. However, instead of using these ideas to describe groups we already know about, we shall in this section see how we can *define new* groups by specifying a set of generators and relations between these generators.

The process is carried out in two stages. Firstly, we consider groups without relations between the generators. Secondly, we consider quotients of such groups.

We shall use the concept of 'words' in the generators in much the same way as we did in Sections 2 and 3 of this unit.

You should now listen to the audio programme for this unit, referring to the tape frames below when asked to do so during the programme.

1 D_6, the symmetry group of the regular hexagon

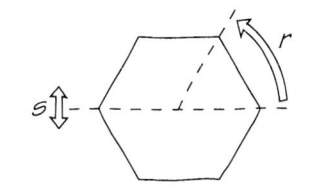

$D_6 = \{e = r^0, r^1, r^2, r^3, r^4, r^5, s, rs, r^2s, r^3s, r^4s, r^5s\}$

Generators: r, s

Relations: $r^6 = e$, $s^2 = e$, $sr = r^5s$

2 A group with 2 generators and 3 relations

Given x, y

Wanted   Group G
generated by x, y
with $x^6 = e$, $y^2 = e$, $yx = x^5y$

Question   Is G isomorphic to D_6?

3 A simple example

One generator x, no relations

4 Consequences of the closure axiom

Must have elements $\{x, xx, xxx, xxxx, ...\}$

*The operation is **concatenation***
For the string xxx we don't distinguish between $(xx)x$ and $x(xx)$

Shorthand: $\{x, x^2, x^3, x^4, ...\}$

5 Consequences of the identity axiom

Must have $xe = x$

Define e = empty string

$x(\ \ ) = x$

Shorthand: $e = x^0$

⑥ Consequences of the inverses axiom

Call the inverse of x the element x^{-1}

Must have $x^{-1}x = xx^{-1} = e$

> x^{-1} cannot be in $\{x^0, x^1, x^2, \ldots\}$

⑦ Modified operation

Operation = concatenation + reduction

Reduction rule:

in any string, replace xx^{-1} and $x^{-1}x$ by the empty string,

e.g. $x^2x^{-1} = xxx^{-1} = xe = x$

⑧ Closure reconsidered

For closure, must augment the set with

$x^{-1}x^{-1}, x^{-1}x^{-1}x^{-1}, \ldots$

Shorthand: $\underbrace{x^{-1}\ldots x^{-1}}_{n \text{ times}} = x^{-n}$

⑨ The free group on one generator

Generator: x

Relations: none

Set: $G = \{\ldots, x^{-2}, x^{-1}, x^0, x^1, x^2, \ldots\}$

Operation: concatenation + reduction

⑩ Checking the closure axiom

Want $\quad x^m x^n \in G = \{\ldots, x^{-2}, x^{-1}, x^0, x^1, x^2, \ldots\}$

$\underline{m, n > 0}\quad x^m x^n = x^{m+n} \in G$

> string of $m+n$ xs

$\underline{m > 0, n < 0}\quad x^m x^n = x^m x^{-k},\ k > 0$

$m > k$: $x^m x^n = \underbrace{x\ldots x}_{m-k} = x^{m-k} = x^{m+n}$

$m < k$: $x^m x^n = \underbrace{x^{-1}\ldots x^{-1}}_{k-m} = x^{-(k-m)} = x^{-k+m} = x^{n+m} = x^{m+n}$

$m = k$: $x^m x^n = x^0 = x^{m-k} = x^{m+n}$

11

Exercise 5.1

Check closure for $x^m x^n$ when

(a) $m, n < 0$

(b) $m < 0, n > 0$

> The cases when m or n are zero are dealt with in Frame 13

12

Solution 5.1

(a) $\underline{m, n < 0}$ $\quad x^m x^n = x^{-j} x^{-k}, \; j, k > 0$

> string of $j+k$ x^{-1}s

$$= x^{-(j+k)} = x^{-j-k} = x^{m+n}$$

(b) $\underline{m < 0, n > 0}$ $\quad x^m x^n = x^{-j} x^n, \; j > 0$

$j > n$: $\; x^m x^n = \underbrace{x^{-1} \ldots x^{-1}}_{j-n} = x^{-(j-n)} = x^{-j+n} = x^{m+n}$

$j < n$: $\; x^m x^n = \underbrace{x \ldots x}_{n-j} = x^{n-j} = x^{n+m} = x^{m+n}$

$j = n$: $\; x^m x^n = x^0 = x^{-j+n} = x^{m+n}$

13

Checking the identity axiom

$x^0 x^n = (\text{empty string}) x^n = x^n$

$x^n x^0 = x^n (\text{empty string}) = x^n$

> Shorthand:
> $x^0 x^n = x^{0+n}$
> $x^n x^0 = x^{n+0}$

14

Rule of indices for composition

Frames 10, 12, 13 show that

$x^m x^n = x^{m+n}$ for all $m, n \in \mathbb{Z}$

15

Checking the associativity axiom

$$\begin{aligned} x^m(x^n x^p) &= x^m x^{n+p} && \text{(rule of indices)} \\ &= x^{m+(n+p)} && \text{(rule of indices)} \\ &= x^{(m+n)+p} && \text{(associativity in } \mathbb{Z}\text{)} \\ &= (x^{m+n}) x^p && \text{(rule of indices)} \\ &= (x^m x^n) x^p && \text{(rule of indices)} \end{aligned}$$

16 *Checking the inverses axiom*

$$x^m x^{-m} = x^{-m} x^m = x^0$$

(both reduce to the empty string)

---

17 *No relations*

Could we have $x^m = x^n$?

If so, $x^m x^{-n} = x^n x^{-n} = x^0$

i.e. x^{m-n} = empty string $\Rightarrow m - n = 0 \Rightarrow m = n$

(Result: the **free group on one generator**)

---

18 *Exercise 5.2*

Prove that the free group on one generator is isomorphic to the integers $\mathbb{Z}$ under addition

---

19 *Solution 5.2*

Let G be the free group with generator x

Consider $f: G \to \mathbb{Z}$
$\quad\quad\quad\quad\quad x^m \mapsto m$ (also possible $x^m \mapsto -m$)

Homomorphism
$f(x^m x^n) = f(x^{m+n}) = m + n = f(x^m) + f(x^n)$

One – one
$f(x^m) = f(x^n) \Rightarrow m = n \Rightarrow x^m = x^n$

Onto
$m \in \mathbb{Z}, f(x^m) = m$

---

20 *The free group on two generators*

Generators:	x, y
Inverses of generators:	x^{-1}, y^{-1}
Words:	sequences of xs, ys, x^{-1}s, y^{-1}s
Elements:	all 'reduced' words
Operation:	concatenation + reduction

21

Examples

Word: $xxy^{-1}yyx^{-1}x^{-1}x^{-1}xyyy$

Reduced: $xxyx^{-1}x^{-1}yyy$

Shorthand: $x^2yx^{-2}y^3$

Product: $(x^2y^{-1}x^3)(x^{-2}yx) = x^2y^{-1}xyx$

22

The free group on n generators: $\mathbf{F}_n$

Generators: $x_1, \ldots, x_n$

Inverses: $x_1^{-1}, \ldots, x_n^{-1}$

Elements: reduced words in $x_1, \ldots, x_n, x_1^{-1}, \ldots, x_n^{-1}$

Operation: concatenation + reduction

23

Introducing relations

$\mathbf{F}_2$ *free group on two generators*

Generators: x, y

Relations: none

D_6 *dihedral group of order 12*

Generators: r, s

Relations: $r^6 = e,\ s^2 = e,\ sr = r^5s$

Standard form for relations: $r^6 = e,\ s^2 = e,\ srsr = e$

$sr = r^5s$ so $sr(r^5s)^{-1} = e$
$srs^{-1}r^{-5} = e$
$srsr = e$

24

The natural homomorphism $\phi: \mathbf{F}_2 \to D_6$

Definition of ϕ

In each element of $\mathbf{F}_2$ (each reduced word), replace

x by r, x^{-1} by r^{-1}

y by s, y^{-1} by s^{-1}

and then calculate the resulting product in D_6

Results

(a) ϕ is a homomorphism

(b) ϕ is onto

(c) Image $D_6 \cong \mathbf{F}_2/\text{Ker}(\phi)$ (by Theorem 4.5)

25 — Ker(ø)

$$\emptyset(x^6) = r^6 = e$$
$$\emptyset(y^2) = s^2 = e$$
$$\emptyset(yxyx) = srsr = e$$

Ker(ø) is the smallest normal subgroup of F_2 containing the set $\{x^6, y^2, yx\,yx\}$

called 'the normal subgroup generated by the set'

26 — Generators and relations

Generators: $a_1, \ldots, a_n$

Relations: $w_1 = e, \ldots, w_k = e$

where $w_1, \ldots, w_k$ are reduced words in the a_i

F_n is the free group with generators $x_1, \ldots, x_n$

K is the normal subgroup of F_n generated by the reduced words in $x_1, \ldots, x_n$ corresponding to $w_1, \ldots, w_k$ (constructed by the method given after Theorem 3.4)

The required group is the quotient F_n/K where the cosets Kx_i correspond to the a_i

---

From the audio programme we know that, given a set of generators

$$\{a_1, \ldots, a_n\}$$

and a set of relations

$$\{w_1 = e, \ldots, w_k = e\},$$

where the w_js are reduced words in the a_is, these define a group

$$G = \langle a_1, \ldots, a_n : w_1 = e, \ldots, w_k = e \rangle.$$

Furthermore, this group is *unique* (up to isomorphism).

However, although we have proved the *existence* of the group, our proof has not told us much about it. In fact, there is a theorem (beyond the scope of this course) which states that there is no algorithm to decide even whether the group defined is the trivial group, containing only the identity. There is also no algorithm to decide whether the group is finite.

Fortunately, if the relations include

$$a_i a_j a_i^{-1} a_j^{-1} = e$$

for all possible generators a_i and a_j, then the group is Abelian and we can completely describe the group. We shall take up this particular case in the Groups stream of the course, and shall give a precise algorithm to identify the group.

APPENDIX: DENOTING GROUPS

Many of the groups discussed in Block 1 have been groups of symmetries of plane figures. In general, the symmetry group of a plane figure X is denoted by

$\Gamma(X)$.

The most familiar example is the symmetry group of the rectangle,

$\Gamma(\square)$.

Other examples include the symmetry groups of plain rectangular friezes and tilings, denoted by

E_1 and E_2

respectively, the frieze groups, denoted by

$\Gamma(F_1) = f_1 = p111$
$\Gamma(F_2) = f_v = pm11$
$\Gamma(F_3) = f_h = p1m1$
$\Gamma(F_4) = f_g = p1a1$
$\Gamma(F_5) = f_r = p112$
$\Gamma(F_6) = f_{vh} = pmm2$ $\qquad$ $\Gamma(F_6)$ is the same group as E_1.
$\Gamma(F_7) = f_{vg} = pma2$

and the dihedral groups

D_n, $\qquad$ Only D_3, D_4 and D_6 have been discussed in Block 1.

where D_n is the group (of order $2n$) of symmetries of the regular n-gon. Each D_n is generated by a rotation r through $2\pi/n$ (so that $r^n = e$) and a reflection s in an axis of symmetry (so that $s^2 = e$). In general we have

$D_n = \langle r, s : r^n = e,\ s^2 = e,\ sr = r^{n-1}s \rangle$.

Notice that the general definition works not only for n-gons with $n \geq 3$ but also for a 2-gon (a line) and a 1-gon (a point), where

$D_2 = \langle r, s : r^2 = e,\ s^2 = e,\ sr = rs \rangle$

and

$D_1 = \langle s : s^2 = e \rangle$. $\qquad$ Here the definition gives $r = e$, r being a rotation through 2π.

Other groups that we have encountered are the Klein group

V $\qquad$ $V \cong \Gamma(\square)$.

and the permutation groups

S_n,

where S_n is the group of permutations of the set $\{1, \ldots, n\}$.

We have also looked at cyclic groups based on the integers:

$\mathbb{Z}$ the integers under addition;
$\mathbb{Z}_n$ the integers $\{0, 1, \cdots, n-1\}$ under addition modulo n;
$\mathbb{Z}_p^*$ the non-zero integers $\{1, \cdots, p-1\}$ under multiplication modulo p, where p is a prime number;

and cyclic groups generated by rotations,

$C_n = \langle r : r^n = e \rangle$,

where r is a rotation through $2\pi/n$.

We have also identified a number of isomorphisms between the various groups:

$\Gamma(\square) \cong V$

$\Gamma(F_1) \cong \Gamma(F_4)$

$\Gamma(F_2) \cong \Gamma(F_5) \cong \Gamma(F_7)$

$\Gamma(F_6) \cong E_1$

$D_3 \cong S_3$

$\mathbb{Z}_4 \cong \mathbb{Z}_5^* \cong C_4$ C_4 is the same group as the subgroup R of D_4 identified in

$\mathbb{Z}_n \cong C_n$ Section 4 of this unit.

There are other isomorphisms between the various groups. We allow you to discover these for yourself.

SOLUTIONS TO THE EXERCISES

Solution 1.1

Using the hint, suppose that x and y are both inverses for a. Then:

$$e = xa \quad \text{(by the inverses axiom)};$$
$$ey = (xa)y \quad \text{(multiplying on the right by } y\text{)}$$
$$= x(ay) \quad \text{(by the associativity axiom)}$$
$$= xe \quad \text{(by the inverses axiom, since } y \text{ is an inverse for } a\text{)}.$$

Therefore, by the identity axiom,

$$y = x.$$

So x and y are the same element and therefore the element a has a unique inverse.

Solution 1.2

The five ways of calculating the product are:

$$w(x(yz)), \quad w((xy)z), \quad (w(xy))z, \quad ((wx)y)z \quad \text{and} \quad (wx)(yz).$$

Solution 1.3

We start where we left off with $(w(xy))z$.

Using the associativity axiom on w, x and y gives

$$(w(xy))z = ((wx)y)z.$$

Using the associativity axiom on (wx), y and z gives

$$((wx)y)z = (wx)(yz).$$

Thus all five products are equal.

Solution 1.4

Right cancellation rule

Let G be a group and let x, y and a be elements of G. Then

$$xa = ya \quad \text{implies} \quad x = y.$$

Proof

Since G is a group and a is an element of G, by the inverses axiom it has an inverse element a^{-1} in G.

We are given that $xa = ya$. Thus, when we multiply both sides of this equation on the right by a^{-1}, we obtain

$$(xa)a^{-1} = (ya)a^{-1}.$$

Using the associativity axiom gives

$$x(aa^{-1}) = y(aa^{-1}).$$

By the inverses axiom this becomes

$$xe = ye,$$

where e is the identity of the group.

Lastly, by the identity axiom, we have the required result that

$$x = y. \qquad \blacksquare$$

Solution 1.5

To prove the result using the uniqueness of inverse elements, we need to verify that $y^{-1}x^{-1}$ has the defining property of the inverse of xy. In other words we need to show that

$$(xy)(y^{-1}x^{-1}) = e = (y^{-1}x^{-1})(xy).$$

We have already seen that $(xy)(y^{-1}x^{-1}) = e$, so it certainly behaves like a *right* inverse for xy. But also

$$\begin{aligned}(y^{-1}x^{-1})(xy) &= y^{-1}((x^{-1}x)y) &&\text{(why?)} \\ &= y^{-1}(ey) &&\text{(why?)} \\ &= y^{-1}y &&\text{(why?)} \\ &= e &&\text{(why?).}\end{aligned}$$

We have not given the justification for each step. We hope that you can, by reference to the proof of Theorem 1.5 if necessary.

So $y^{-1}x^{-1}$ also behaves as a *left* inverse for xy.

Hence, by the uniqueness of inverses, $y^{-1}x^{-1}$ is the inverse of xy.

Solution 1.6

We give a proof based on cancellation.

Because $\left(x^{-1}\right)^{-1}$ is the inverse of x^{-1}, we have

$$x^{-1}\left(x^{-1}\right)^{-1} = e.$$

On the other hand, x^{-1} is the inverse of x, and so

$$x^{-1}x = e.$$

Equating the two expressions for e gives

$$x^{-1}\left(x^{-1}\right)^{-1} = x^{-1}x,$$

and left cancellation gives the desired result.

Solution 1.7

We are given that $yx = e$ and, by the inverses axiom, we know that $x^{-1}x = e$. It follows that

$$yx = x^{-1}x \, (= e).$$

Using the Right Cancellation Rule, we have the required result that $y = x^{-1}$.

An alternative proof involves multiplying both sides of the equation $yx = e$ on the right by the element x^{-1}.

Solution 1.8

(a) We are given that $fx = x$ and, by the identity axiom, we know that $ex = x$.

Hence $fx = ex$ and, by the Right Cancellation Rule, $f = e$.

An alternative proof involves multiplying both sides of the equation $fx = x$ on the right by the element x^{-1}.

(b) The corresponding statement to the above result is the following.

> Let G be a group with identity e and let x be an element of G.
> If f is an element of G such that $xf = x$, then $f = e$.
> In other words, if f behaves like a right identity for the particular element x of G, then f *is* the identity of the group.

Proof

We are given that $xf = x$ and, by the identity axiom, we know that $xe = x$.

Hence $xf = xe$ and, by the Left Cancellation Rule, $f = e$. ∎

An alternative proof involves multiplying both sides of the equation $xf = xe$ on the left by the element x^{-1}.

Solution 1.9

The underlying set is

$$\mathbb{Z}_2 \times \mathbb{Z}_3 = \{0,1\} \times \{0,1,2\}$$
$$= \{(0,0),(0,1),(0,2),(1,0),(1,1),(1,2)\}.$$

The Cayley table is constructed using addition modulo 2 for the first components and addition modulo 3 for the second.

+	(0,0)	(0,1)	(0,2)	(1,0)	(1,1)	(1,2)
(0,0)	(0,0)	(0,1)	(0,2)	(1,0)	(1,1)	(1,2)
(0,1)	(0,1)	(0,2)	(0,0)	(1,1)	(1,2)	(1,0)
(0,2)	(0,2)	(0,0)	(0,1)	(1,2)	(1,0)	(1,1)
(1,0)	(1,0)	(1,1)	(1,2)	(0,0)	(0,1)	(0,2)
(1,1)	(1,1)	(1,2)	(1,0)	(0,1)	(0,2)	(0,0)
(1,2)	(1,2)	(1,0)	(1,1)	(0,2)	(0,0)	(0,1)

Solution 1.10

Closure Let (x_1, y_1) and (x_1, y_2) be any two elements of $\mathbb{R} \times \mathbb{R}$. By definition, x_1, y_1, x_2 and y_2 are in $\mathbb{R}$. Since $\mathbb{R}$ is closed under addition, it follows that $x_1 + x_2$ and $y_1 + y_2$ are also in $\mathbb{R}$. By definition,

$$(x_1, y_1) + (x_2, y_2) = (x_1 + x_2, y_1 + y_2),$$

which, by the above, is an element of $\mathbb{R} \times \mathbb{R}$. Therefore the operation $+$ satisfies the closure axiom.

Associativity Let (x_1, y_1), (x_2, y_2) and (x_3, y_3) be three elements of $\mathbb{R} \times \mathbb{R}$. By definition,

$$((x_1, y_1) + (x_2, y_2)) + (x_3, y_3) = ((x_1 + x_2) + x_3, (y_1 + y_2) + y_3)$$

and

$$(x_1, y_1) + ((x_2, y_2) + (x_3, y_3)) = (x_1 + (x_2 + x_3), y_1 + (y_2 + y_3)).$$

The right-hand sides of these two equations are equal, since addition in $\mathbb{R}$ is associative. Hence

$$((x_1, y_1) + (x_2, y_2)) + (x_3, y_3) = (x_1, y_1) + ((x_2, y_2) + (x_3, y_3))$$

and so $+$ satisfies the associativity axiom.

Identity The element $(0,0)$ belongs to $\mathbb{R} \times \mathbb{R}$, and

$$(0,0) + (x,y) = (x,y) + (0,0) = (x,y)$$

for any (x,y) in $\mathbb{R} \times \mathbb{R}$. Therefore $(0,0)$ is the identity of $\mathbb{R} \times \mathbb{R}$, and the identity axiom is satisfied.

Inverses If (x,y) is any element of $\mathbb{R} \times \mathbb{R}$ then, as x and y are in $\mathbb{R}$, their additive inverses $-x$ and $-y$ are also in $\mathbb{R}$. So the element $(-x, -y)$ is an element of $\mathbb{R} \times \mathbb{R}$. Furthermore,

$$(x, y) + (-x, -y) = (0, 0) = (-x, -y) + (x, y).$$

So $(-x, -y)$ is the inverse of the element (x, y), and the inverses axiom is satisfied.

Solution 1.11

(a) Let (g_1, h_1), (g_2, h_2) and (g_3, h_3) be any three elements of $G \times H$.

$$\begin{aligned}
((g_1, h_1) \square (g_2, h_2)) \square (g_3, h_3) &= (g_1 \circ g_2, h_1 * h_2) \square (g_3, h_3) && \text{(by the definition of } \square) \\
&= ((g_1 \circ g_2) \circ g_3, (h_1 * h_2) * h_3) && \text{(by the definition of } \square) \\
&= (g_1 \circ (g_2 \circ g_3), h_1 * (h_2 * h_3)) && \text{(by the associativity axiom)} \\
&= (g_1, h_1) \square ((g_2 \circ g_3), (h_2 * h_3)) && \text{(by the definition of } \square) \\
&= (g_1, h_1) \square ((g_2, h_2) \square (g_3, h_3)) && \text{(by the definition of } \square).
\end{aligned}$$

(b) Let (g, h) be any element of the direct product $G \times H$. Then

$$\begin{aligned}
(e, f) \square (g, h) &= (e \circ g, f * h) && \text{(by the definition of } \square) \\
&= (g, h) && \text{(by the identity axiom)}.
\end{aligned}$$

Similarly

$$(g, h) \square (e, f) = (g, h).$$

Hence (e, f) is the identity element of the direct product.

(c) Applying the inverse properties from the individual groups,

$$\begin{aligned}
(g, h) \square (g^{-1}, h^{-1}) &= (g \circ g^{-1}, h * h^{-1}) && \text{(by the definition of } \square) \\
&= (e, f) && \text{(by the inverses axiom)}.
\end{aligned}$$

Similarly

$$(g^{-1}, h^{-1}) \square (g, h) = (e, f).$$

Hence (g^{-1}, h^{-1}) is the inverse of (g, h).

Solution 2.1

Closure The only products not involving the identity are as follows:

$$(123)(123) = (132)$$
$$(123)(132) = e$$
$$(132)(123) = e$$
$$(132)(132) = (123)$$

Hence the subset satisfies the closure axiom.

Identity The element e is in the subset, so the identity axiom is satisfied.

Inverses

$$e^{-1} = e$$
$$(123)^{-1} = (132)$$
$$(132)^{-1} = (123)$$

So the subset satisfies the inverses axiom.

Hence

$$\{e, (123), (132)\} \leq S_3.$$

Solution 2.2

We are given that H is a subgroup of the group G, so H satisfies the subgroup axioms. In particular, it is a subset of G.

Since it satisfies the identity axiom, H contains the identity element of G and so is non-empty.
So H satisfies the first property.

Now suppose that x and y are any two elements of H.
Since H satisfies the inverses axiom, and x is an element of H, so is the element x^{-1}.
We now know that x^{-1} and y are elements of H. Being a subgroup, H satisfies the closure axiom, so the element $x^{-1}y$ is in H.
So H satisfies the second property.

Solution 2.3

If

We are given that H is a non-empty subset of G and that if x and y are in G then xy^{-1} is in G. We have to show that H is a subgroup of G; in other words, that it satisfies the three subgroup axioms. We deal with them in the same order as before.

Identity

By the first property, H is non-empty. Therefore there exists some element x in H.
By the second property, x and y in H imply that xy^{-1} is in H. Taking the special case $y = x$ shows that $xx^{-1} = e$ is in H.
Thus H satisfies the identity axiom.

Inverses

If y is in H then, by the identity axiom (already verified), both e and y are in H.
It follows therefore, by the second property, that $ey^{-1} = y^{-1}$ is in H.
So H satisfies the inverses axiom.

Closure

Let x and y be elements of H. By the inverses axiom (already verified), y^{-1} is an element of H. Therefore x and y^{-1} are both in H.
So, by the second property, $x\left(y^{-1}\right)^{-1} = xy$ is in H.
Therefore H satisfies the closure axiom.

Only if

We are given that H is a subgroup of the group G, so it satisfies the subgroup axioms. In particular, it is a subset of G.

Since it satisfies the identity axiom, H contains the identity element of G and so is non-empty.
So H satisfies the first property.

Now suppose that x and y are any two elements of H.
Since H satisfies the inverses axiom, and y is an element of H, so is the element y^{-1}.
We now know that x and y^{-1} are elements of H. Being a subgroup, H satisfies the closure axiom, so the element xy^{-1} is in H.
So H satisfies the second property.

Solution 2.4

The group $\mathbb{Z}_6$ under addition modulo 6 has subgroups $\{0, 3\}$ and $\{0, 2, 4\}$, and the union of these is $\{0, 2, 3, 4\}$.

This set is not a subgroup since it is not closed under addition modulo 6. For example, $2 + 3 = 5 \pmod 6$ is not in the set.

An alternative reason why the set is not a subgroup is provided by Lagrange's Theorem, since the subset has 4 elements and 4 does not divide 6.

Solution 2.5

We have

$$sr^{2k} = s\overbrace{r^2\ldots r^2}^{k \text{ copies}}$$
$$= r^5 s \overbrace{r^2\ldots r^2}^{k-1 \text{ copies}} \quad (\text{using } sr = r^5 s)$$
$$= r^5 r^5 s \overbrace{r^2\ldots r^2}^{k-1 \text{ copies}} \quad (\text{using } sr = r^5 s)$$
$$= r^{10} s \overbrace{r^2\ldots r^2}^{k-1 \text{ copies}}$$
$$= r^{20} s \overbrace{r^2\ldots r^2}^{k-2 \text{ copies}} \quad (\text{using } sr = r^5 s \text{ twice})$$
$$\ldots$$
$$= r^{10k} s$$
$$= r^{2\times 5k} s$$
$$= r^{2l} s,$$

where $l = 5k$.

Since we have the relation $r^6 = e$, the only distinct even powers $2l$ of r are $r^0 = e$, r^2 and r^4, corresponding to $l = 0, 1, 2$.

An alternative method of showing that $sr^{2k} = r^{2l}s$, where $l = 5k$, is to make use of Theorem 1.1 of *Unit IB3* to deduce that
$$sr^{2k} = r^{5(2k)}s = r^{2(5k)}s.$$

Solution 2.6

$$e\{e, (12)\} = \{e, (12)\};$$
$$(12)\{e, (12)\} = \{(12), (12)(12)\}$$
$$= \{(12), e\};$$
$$(13)\{e, (12)\} = \{(13), (13)(12)\}$$
$$= \{(13), (123)\};$$
$$(23)\{e, (12)\} = \{(23), (23)(12)\}$$
$$= \{(23), (132)\};$$
$$(123)\{e, (12)\} = \{(123), (123)(12)\}$$
$$= \{(123), (13)\};$$
$$(132)\{e, (12)\} = \{(132), (132)(12)\}$$
$$= \{(132), (23)\}.$$

Thus there are just three distinct left cosets, namely

$$\{e, (12)\}, \quad \{(13), (123)\} \quad \text{and} \quad \{(23), (132)\}.$$

Solution 2.7

Taking left cosets with e, r^2 and r^4 gives:
$$e\{e,r^2,r^4\} = \{e,r^2,r^4\} = H;$$
$$r^2\{e,r^2,r^4\} = \{r^2,r^4,r^6\}$$
$$= \{r^2,r^4,e\} = H;$$
$$r^4\{e,r^2,r^4\} = \{r^4,r^6,r^8\}$$
$$= \{r^4,e,r^2\} = H.$$

Taking left cosets with elements of D_6 not in H gives:
$$r\{e,r^2,r^4\} = \{r,r^3,r^5\} \neq H;$$
$$r^3\{e,r^2,r^4\} = \{r^3,r^5,r^7\}$$
$$= \{r^3,r^5,r\} \neq H;$$
$$r^5\{e,r^2,r^4\} = \{r^5,r^7,r^9\}$$
$$= \{r^5,r,r^3\} \neq H;$$
$$s\{e,r^2,r^4\} = \{s,sr^2,sr^4\}$$
$$= \{s,r^4s,r^2s\} \neq H;$$
$$rs\{e,r^2,r^4\} = \{rs,rsr^2,rsr^4\}$$
$$= \{rs,r^5s,r^3s\} \neq H;$$
$$r^2s\{e,r^2,r^4\} = \{r^2s,r^2sr^2,r^2sr^4\}$$
$$= \{r^2s,s,r^4s\} \neq H;$$
$$r^3s\{e,r^2,r^4\} = \{r^3s,r^3sr^2,r^3sr^4\}$$
$$= \{r^3s,rs,r^5s\} \neq H;$$
$$r^4s\{e,r^2,r^4\} = \{r^4s,r^4sr^2,r^4sr^4\}$$
$$= \{r^4s,r^2s,s\} \neq H;$$
$$r^5s\{e,r^2,r^4\} = \{r^5s,r^5sr^2,r^5sr^4\}$$
$$= \{r^5s,r^3s,rs\} \neq H.$$

Alternatively, and more briefly, if x is any element of D_6 not in H, then
$$x\{e,r^2,r^4\} = \{x,xr^2,xr^4\} \neq H,$$
since $x \notin H$.

Thus the conjecture is verified.

Solution 2.8

We are given that $aH = bH$ and must show that $a^{-1}b \in H$.

Since $b = be$, the element b is in the coset bH. But $aH = bH$, so $b \in aH$. Therefore there exists some $h \in H$ such that $b = ah$.

So $a^{-1}b = a^{-1}ah = h$, and therefore $a^{-1}b$ is in H.

Solution 2.9

(a) **One–one** If $f(x) = f(y)$ then, by the definition of f, we know that
$$ax = ay.$$
So, by the Left Cancellation Rule, $x = y$. Hence f is one–one.

Onto If z is any element in the codomain aH, then $z = ah$ for some $h \in H$. Now, by definition, $f(h) = ah$ and therefore $f(h) = z$. Hence f is onto.

(b) By part (a), the left cosets aH and bH each have the same number of elements as the subgroup H.

Solution 3.1

Well-defined Using the hint, suppose that $aH = bH$. We must show that $\phi(aH) = \phi(bH)$, in other words that
$$Ha^{-1} = Hb^{-1}.$$

Now, by Theorem 2.6, we have
$$aH = bH \quad \text{if and only if} \quad a^{-1}b \in H$$
and, by Theorem 3.1, we have
$$Ha^{-1} = Hb^{-1} \quad \text{if and only if} \quad a^{-1}(b^{-1})^{-1} \in H.$$

But
$$a^{-1}(b^{-1})^{-1} = a^{-1}b \in H,$$
since $aH = bH$. Therefore $Ha^{-1} = Hb^{-1}$ and ϕ is well-defined.

One–one Suppose that $\phi(aH) = \phi(bH)$, i.e. $Ha^{-1} = Hb^{-1}$. It follows, from Theorem 3.1, that
$$a^{-1}(b^{-1})^{-1} = a^{-1}b \in H.$$

Therefore, by Theorem 2.6, $aH = bH$ and ϕ is one–one.

Onto Suppose that Hx is any right coset of H in G. Then $x^{-1} \in G$ and
$$\phi(x^{-1}H) = H(x^{-1})^{-1} = Hx,$$
and so ϕ is onto.

Solution 3.2

We already know, from *Unit IB2*, that both $\{e\}$ and G are subgroups of G.

If a is any element of G, then
$$a\{e\} = \{e\}a = \{a\}.$$

Hence $\{e\}$ is a normal subgroup of G.

To show that G is a normal subgroup of G, we show that if a is any element of G, then aG and Ga are both equal to G.

Since $a \in G$, by Lemma 2.1, $aG = G$. By the corresponding result for right cosets, $Ga = G$. So for all $a \in G$, we have $aG = Ga = G$, and hence G is a normal subgroup of G.

Solution 3.3

We have already seen in Section 2 that H is a subgroup of S_3.

Calculating the left and right cosets for H gives the following:
$$eH = He = H;$$
$$(12)H = H(12) = \{(12),(23),(13)\};$$
$$(13)H = H(13) = \{(13),(12),(23)\};$$
$$(23)H = H(23) = \{(23),(13),(12)\};$$
$$(123)H = H(123) = \{(123),(132),e\};$$
$$(132)H = H(132) = \{(132),e,(123)\}.$$

Hence each left coset is equal to the corresponding right coset, and so H is a normal subgroup.

Solution 3.4

By the definition of index, H has exactly two left cosets and exactly two right cosets in G. Suppose aH is a left coset. There are two possibilities, $a \in H$ or $a \notin H$.

If $a \in H$, then, by Lemma 2.1 and the corresponding result for right cosets, $aH = Ha = H$.

If $a \notin H$, then, by Lemma 2.1 and the corresponding result for right cosets, $aH \neq H$ and $Ha \neq H$. Since G has two left cosets and two right cosets, it follows, from Theorems 2.7 and 3.2, that aH and Ha are each equal to the set of the elements of G which are not in $H = eH = He$.

Hence $aH = Ha$ and H is a normal subgroup of G.

Solution 3.5

We are given that $aHa^{-1} \subseteq H$ and want to show that Ha is a subset of aH.

Let x be in Ha so that $x = h_1 a$ for some h_1 in H. We wish to show that x is in aH, i.e. that $x = ah$ for some $h \in H$. So we need to show that $a^{-1}x \in H$.

Now $a^{-1}x = a^{-1}h_1 a$, and if we rewrite this as

$$a^{-1}h_1 a = a^{-1}h_1(a^{-1})^{-1}$$

it shows that $a^{-1}x$ is an element of the set $a^{-1}H(a^{-1})^{-1}$.

Now the given condition says that

$$(\text{element of } G) H (\text{element of } G)^{-1} \subseteq H$$

for all elements of G. Using the element a^{-1} shows that

$$a^{-1}H(a^{-1})^{-1} = a^{-1}Ha \subseteq H.$$

Therefore $a^{-1}x = h_2$ for some $h_2 \in H$.

Hence $x = ah_2$, which is an element of aH, and so Ha is a subset of aH.

Solution 3.6

Let N be the intersection of the given collection of normal subgroups of G. As we have observed, N is a subgroup of G.

Now suppose that H is any normal subgroup in the collection. Since $N \subseteq H$,

$$aNa^{-1} \subseteq aHa^{-1}$$
$$\subseteq H \quad \text{(by Theorem 3.3).}$$

This argument shows that aNa^{-1} is contained in *every* subgroup of the collection, and hence is contained in their intersection. That is,

$$aNa^{-1} \subseteq N,$$

and so N is a normal subgroup of G (by Theorem 3.3).

Solution 3.7

If x is in $\tilde{S}$, then $x = asa^{-1}$ for some s in $\hat{S}$. So

$$x^{-1} = \left(asa^{-1}\right)^{-1} = \left(a^{-1}\right)^{-1} s^{-1} a^{-1} = as^{-1}a^{-1}.$$

Since $s^{-1} \in \hat{S}$, it follows that

$$x^{-1} = as^{-1}a^{-1} \in \tilde{S}.$$

Solution 3.8

Using the definition of the product of two cosets, and the associativity axiom for the group G, we have

$$aN(bNcN) = aN(bcN)$$
$$= a(bc)N$$
$$= (ab)cN$$
$$= (ab)NcN$$
$$= (aNbN)cN.$$

Solution 3.9

As has been previously remarked, $N = eN$. So, for any element $aN \in G/N$, we have

$$NaN = eNaN = (ea)N = aN.$$

Similarly, $aNN = aN$.

Hence N is the identity element of G/N.

Solution 3.10

By the definition of the product in G/N,

$$a^{-1}NaN = (a^{-1}a)N = eN = N.$$

Similarly,

$$aNa^{-1}N = (aa^{-1})N = eN = N.$$

Hence $a^{-1}N$ is the inverse of aN.

Solution 3.11

(a) Firstly we show that N is a subgroup of G.

Closure The only non-trivial product to check is $r^2r^2 = e$. So N satisfies the closure axiom.

Identity As e is in N, the identity axiom is satisfied.

Inverses As $e^{-1} = e$ and $(r^2)^{-1} = r^2$, the inverses axiom is satisfied.

Hence N is a subgroup of G.

Next, by Theorem 3.3, to show that N is normal, we need to check that, for each a in G, we have that aNa^{-1} is a subset of N.

For each a in G, $aea^{-1} = e \in N$, so all that we have to check is that ar^2a^{-1} is in N. We have, using the relations $r^4 = s^2 = e$ and $sr = r^3s$:

$$er^2e^{-1} = r^2$$
$$rr^2r^{-1} = r^2$$
$$(r^2)r^2(r^2)^{-1} = r^2$$
$$(r^3)r^2(r^3)^{-1} = r^2$$
$$sr^2s^{-1} = r^2$$
$$(rs)r^2(rs)^{-1} = r^2$$
$$(r^2s)r^2(r^2s)^{-1} = r^2$$
$$(r^3s)r^2(r^3s)^{-1} = r^2$$

This completes the proof that N is a normal subgroup of G.

(b) The distinct left cosets of $N = \{e, r^2\}$ are:

$$eN = r^2N = N = E$$
$$rN = r^3N = \{r, r^3\} = A$$
$$sN = r^2sN = \{s, r^2s\} = B$$
$$rsN = r^3sN = \{rs, r^3s\} = C$$

(c) The Cayley table of G/N is:

∘	E	A	B	C
E	E	A	B	C
A	A	E	C	B
B	B	C	E	A
C	C	B	A	E

Solution 3.12

As usual, we need to prove two set inclusions. For convenience, let

$$P = \{xy : x \in aN, y \in bN\},$$
$$Q = \{(ab)n : n \in N\}.$$

Firstly we show that $P \subseteq Q$.
Let z be an element of P. Then $z = an_1 bn_2$ for some n_1 and n_2 in N.
Now $n_1 b$ is an element of Nb and, as N is a normal subgroup, $Nb = bN$.
Hence $n_1 b = bn_3$ for some n_3 in N.
Therefore $z = an_1 bn_2 = abn_3 n_2$, which is an element of Q.
This shows that $P \subseteq Q$.

Now we show that $Q \subseteq P$.
Let z be an element of Q. So $z = abn$ for some n in N.
But a is in aN and bn is in bN. Therefore $z = (a)(bn)$ is in P.
This shows that $Q \subseteq P$.

Hence $P = Q$.

Solution 4.1

The Cayley table of the group $(\mathbb{Z}_4, +_4)$ is:

$+_4$	0	1	2	3
0	0	1	2	3
1	1	2	3	0
2	2	3	0	1
3	3	0	1	2

Solution 4.2

If $R = \langle r \rangle$, then R is the cyclic subgroup of D_4 generated by r. Since $r^4 = e$,

$$\langle r \rangle = \{e, r, r^2, r^3\}.$$

Its Cayley table is:

∘	e	r	r^2	r^3
e	e	r	r^2	r^3
r	r	r^2	r^3	e
r^2	r^2	r^3	e	r
r^3	r^3	e	r	r^2

Solution 4.3

Clearly ϕ is a function from $\mathbb{Z}$ to $2\mathbb{Z}$.

One–one If $\phi(a) = \phi(b)$, then $2a = 2b$ and so $a = b$, proving that ϕ is one–one.

Onto If a is in $2\mathbb{Z}$, then $a = 2m$, for some $m \in \mathbb{Z}$. Hence $a = \phi(m)$, proving that ϕ is onto.

Lastly, we check that ϕ preserves the binary operation:

$$\begin{aligned}\phi(a+b) &= 2(a+b) \\ &= 2a + 2b \\ &= \phi(a) + \phi(b).\end{aligned}$$

Solution 4.4

As $aa^{-1} = e$, we have that $\phi(a)\phi(a^{-1}) = \phi(e)$.

But, by Theorem 4.2, $\phi(e)$ is the identity of H, so $\phi(a^{-1})$ behaves like the right inverse of $\phi(a)$ in H.

Therefore, by Theorem 1.6 (or Lemma 1.1), $\phi(a^{-1})$ is the inverse of $\phi(a)$ in H, i.e.
$$(\phi(a))^{-1} = \phi(a^{-1}).$$

Solution 4.5

In the group $\mathbb{Z}_5^*$ we have that $3^2 = 4$, $3^3 = 2$ and $3^4 = 1$. Hence the element 3 is a generator of the group.

Since 1 is a generator of $\mathbb{Z}_4$, we might therefore look for an isomorphism from $\mathbb{Z}_4$ to $\mathbb{Z}_5^*$ for which $\phi(1) = 3$.

The inverse of 1 in $\mathbb{Z}_4$ is 3 and the inverse of 3 in $\mathbb{Z}_5^*$ is 2, so we would also need $\phi(3) = 2$. Since we know that for any isomorphism $\phi(0) = 1$, this leaves $\phi(2) = 4$.

Rewriting the Cayley table as

$\times_5$	1	3	4	2
1	1	3	4	2
3	3	4	2	1
4	4	2	1	3
2	2	1	3	4

we find that it does have the same pattern as the Cayley table for $(\mathbb{Z}_4, +_4)$, and so the function ϕ defined by

$$\phi(0) = 1$$
$$\phi(1) = 3$$
$$\phi(2) = 4$$
$$\phi(3) = 2$$

is an isomorphism from $\mathbb{Z}_4$ to $\mathbb{Z}_5^*$.

Solution 4.6

(a) We wish to show that the group H is Abelian, so we must show that, for any x and y in H, we have $xy = yx$.

Since ϕ is onto, there exist elements a and b in G such that $\phi(a) = x$ and $\phi(b) = y$.

Now G is an Abelian group, so $ab = ba$. Therefore, since ϕ is an isomorphism,
$$\begin{aligned} xy &= \phi(a)\phi(b) \\ &= \phi(ab) \\ &= \phi(ba) \\ &= \phi(b)\phi(a) \\ &= yx. \end{aligned}$$

Hence H is Abelian.

(b) Since $\mathbb{Z}_6$ is an Abelian group and S_3 is not, they cannot be isomorphic.

Solution 4.7

Using the given hint, suppose that a has infinite order but that $\phi(a)$ has finite order n. Then, since ϕ^{-1} is an isomorphism (by Theorem 4.1), $\phi^{-1}(\phi(a)) = a$ also has finite order n (by Theorem 4.3). This is a contradiction, and so $\phi(a)$ must have infinite order.

Solution 4.8

There are several ways of proving this. For example:

$\mathbb{Z}_2 \times \mathbb{Z}_3$ is Abelian whereas S_3 is not;

$\mathbb{Z}_2 \times \mathbb{Z}_3$ has just one element of order 2, namely $(1,0)$, whereas S_3 has three, namely (12), (13) and (23);

$\mathbb{Z}_2 \times \mathbb{Z}_3$ has an element, $(1,1)$, of order 6, whereas the elements of S_3 have orders 1, 2 and 3.

This last remark shows that the group $\mathbb{Z}_2 \times \mathbb{Z}_3$ is cyclic, so you should be able to find an isomorphism between it and $\mathbb{Z}_6$.

Solution 4.9

(a) By definition, ϕ is a function from $\mathbb{Z}$ to $\mathbb{Z}$. Furthermore, for all m and n in the domain $\mathbb{Z}$,

$$\phi(m+n) = 3(m+n) = 3m + 3n = \phi(m) + \phi(n).$$

Hence ϕ has the morphism property and is a homomorphism from $(\mathbb{Z}, +)$ to itself.

However, ϕ is not an isomorphism, since it is not onto; for example, the element 1 of the codomain is not in the image set of ϕ.

(b) The kernel consists of all elements mapping to the identity, 0, of the codomain. Thus we want all $n \in \mathbb{Z}$ such that

$$\phi(n) = 3n = 0.$$

The only possibility is $n = 0$. Thus

$$\text{Ker}(\phi) = \{0\},$$

which is trivially a subgroup of the domain.

(c) The elements of the image are those integers of the form $\phi(n) = 3n$ for some $n \in \mathbb{Z}$. Hence

$$\text{Im}(\phi) = \{3n : n \in \mathbb{Z}\} = 3\mathbb{Z}.$$

Now $3\mathbb{Z}$ is a non-empty set, since $0 \in 3\mathbb{Z}$. Also, if $3m$ and $3n$ are any two elements of $3\mathbb{Z}$, then

$$3m - 3n = 3(m-n) \in 3\mathbb{Z}.$$

Thus $3\mathbb{Z}$ is a subgroup of the codomain $\mathbb{Z}$, by Theorem 2.1.

Recall the additive form of Theorem 2.1 on page 14.

Solution 4.10

We need to check that, for all a and b in G,

$$\psi(aK\,bK) = \psi(aK)\psi(bK).$$

Now,

$$\begin{aligned}\psi(aK\,bK) &= \psi(abK) \\ &= \phi(ab) \\ &= \phi(a)\phi(b) \\ &= \psi(aK)\psi(bK).\end{aligned}$$

OBJECTIVES

After you have studied this unit, you should be able to:

(a) use the group axioms to check whether given sets with binary operations are groups;

(b) form the direct product of two groups;

(c) check whether a given subset of a group is a subgroup;

(d) find the subgroup generated by a set of elements of a group;

(e) partition a group into cosets of a given subgroup;

(f) check whether a given subgroup of a group is normal;

(g) find the normal subgroup generated by a set of elements of a group;

(h) form and identify quotient groups in simple cases;

(i) check whether a given function is an isomorphism or homomorphism;

(j) decide whether two given groups are isomorphic;

(k) find kernels and images of homomorphisms;

(l) construct proofs, similar to those in the unit, based on the group axioms and the results of the unit.

INDEX

associativity 6
Cartesian product 11
closure 6
concetenation 42
coset 19, 23
direct product of groups 11
disjoint sets 21
empty string 42
finite group 22
first isomorphism theorem 40
free group 42
generalized associativity rule 8
generator 42
group 6
group axioms 6
homomorphism 36
homomorphism property 36
identity element 6

image 37
index of subgroup 23
infinite group 22
infinite order of element 35
infinite order of group 22
inverse element 6
inverse of product 9
isomorphism 31
kernel 37
Lagrange's theorem 22
left cancellation rule 8
left coset 19
morphism property 36
natural homomorphism 41
normal subgroup 24
ordered pair 11
order of group 22
order of group element 35

partition 22
product of cosets 27
product of sets 19
quotient group 28
reduction 42
relation 42
right cancellation rule 8, 9, 51
right coset 23
string 42
subgroup 12
subgroup axioms 12
subgroup generated by set 15
underlying set 6
uniqueness of identity 6
uniqueness of inverses 7
well-defined 23
words 18